Predictive Dialing Fundamentals

An overview of predictive dialing technologies, their applications and usage today, with a predominant emphasis on outbound call management

by Aleksander Szlam
Ken Thatcher
Melita International Corporation

Flatiron Publishing, Inc.
New York

Published by Telecom Books
An imprint of Miller Freeman, Inc.
12 West 21st St., N.Y., N.Y., 10010
www.telecombooks.com

ISBN 0-936648-80-5

For individual orders, and for information on special discounts for quantity orders, please contact:
Telecom Books
6600 Silacci Way
Gilroy, CA 95020
Tel:800-LIBRARY or 408-848-3854
Fax:408-848-5784
Email:telecom@rushorder.com

Distributed to the book trade in the U.S. and Canada by
Publishers Group West
1700 Fourth St., Berkeley, CA 94710

Manufactured in the United States of America

First Edition, February 1996
Transferred to Digital Printing 2009

TABLE OF CONTENTS

ACKNOWLEDGMENT

FOREWORD

PREFACE

Predictive Dialing Fundamentals
 New to the Industry
 Already Implementing a Predictive Dialer
 Familiar with Current Technologies

ACKNOWLEDGMENTS

- We would like to thank Jim Owen and Ed Miller for their contributions to Chapter 3 and Chapter 6.

- We would also like to thank Margo Gathright-Dietrich, John Morgan, Shebra Johnson and Andrea Gassner for their editing, indexing, support and advice.

- Very special thanks to Andrea Boyle for preparing and typing all the material, and for working diligently to provide outstanding graphics throughout the book.

- Special thanks to Diane Bates for editing the book.

- Special thanks to Ginny Fisher, who relentlessly pursued getting it all done on time!

- Special thanks to Scott Aiken for the art design of the book cover.

- And finally, our hero, Joel Artzt who had to come up to speed on the topic of predictive dialing so that he could organize and edit all of the copy.

Why I Love Predictive Dialers

by Harry Newton

editor-in-chief, Call Center Magazine

I love predictive dialers because they deliver the benefits they promise. And they do it neatly and painlessly. I've never met a predictive dialer user who didn't realize the productivity improvements he was promised and figured on to justify the system.

I have only one problem with predictive dialers. Most people use them for collecting money. Dial up the deadbeats. Dun them for money. B-O-R-I-N-G!

I'm a mail order junkie. Everyone is. Mail order catalogs have become our escapist reading. I'm dressed in clothes from L.L. Bean, Eddie Bauers, Nordstroms, etc. I'm typing on a computer from PC Connection. My computer is tied into a server. We bought the wire and the server and the network cards by mail order.

But I'm an unloved mail order junkie. No one ever calls me. Bean and Bauers know I like "large tall." They know I'm 11N in shoes. They know what I like and what I don't like. I've been buying from them for years. How come they never call me and tell me of their fantastic tall and narrow new products?

I'm Platinum on American Airlines. Their most frequent flyer. So why don't they call me and offer me a special "weekend" surprise? Call me on Wednesday. Fill their planes on Friday and Sunday. Don't call me if their planes are full.

To my naive brain, predictive dialers are the best sales tool ever invented. They let you sell when you've got something to sell. And not sell when you got nothing to sell. They let you sell "talls" to tall people and narrow shoes to narrowed-footed people. This is not rocket science.

One day, some companies will figure there are millions of people (like me) waiting at the end of their phone line for someone to call them and sell them something. That someone, if they're smart, will be using a predictive dialer.

PREFACE

PREDICTIVE DIALING FUNDAMENTALS

The intent of this book is to provide an overall understanding of predictive dialing technologies, their applications and usage today, with a predominant emphasis on inbound and outbound call management.

Predictive Dialing Fundamentals provides basic information about predictive dialers: what they are, how they work, and how they benefit corporations. It is not intended to provide an in-depth understanding of WorkFlow, IVR/Voice Mail, LAN/WAN, PBX/ACD, CTI technologies, and other call center related technologies, but it will primarily concentrate on inbound/outbound predictive dialing applications which are submerged in these technologies.

Some information in this book may be repetitive. However, it is the intention of the authors to educate any reader, who may indeed select to read only some of the chapters that may be applicable; hence, each chapter stands on its own merits.

NEW TO THE INDUSTRY

If you're a newcomer to predictive dialers, and are responsible for analyzing, recommending, and deploying outbound call center solutions today, with the view of what tomorrow will bring then check out Chapter 1. This chapter provides a comprehensive thorough overview of the entire predictive dialing environment. Then consider Chapter 4, which provides simple formulas to determine if you should use predictive dialers and highlights six case studies and exploring the issues each company experienced that caused them to deploy a predictive dialer. The case studies continue to highlight how different companies in banking/financial, cable, healthcare, utility and services sectors implemented the dialer and benefited from its use. Then, if you're interested in what you need to look for when choosing a predictive dialer, proceed directly to Chapter 5 and further into Chapter 6 for a clear understanding of the deployment and integration process.

ALREADY IMPLEMENTING A PREDICTIVE DIALER

If you're already implementing a predictive dialer then you may want to start with Chapters 3, 4 and 5.

FAMILIAR WITH CURRENT TECHNOLOGIES

While, if you're more familiar with the technologies (i.e., CTI, PBX, ACDs) consider starting with Chapter 3 which covers the technology and in particular the role that client/server plays in the development of call center solutions. Then move on to Chapter 7, which explores the call center evolution and how technologies will continue to support the "Customer is King" philosophy.

The authors of this book have been involved with the invention and development of the predictive dialer industry, and pioneered and invented several of the key underlying technologies, which significantly contributed to and continue to shape the success of many call centers across the world. Some of these innovations are:

- "Hello" - Call Progress Detection
- Outbound Predictive Dialing
- Synchronized "screen pops" based on ANI/DNIS or account information
- Computer Telephony Integration (CTI) of ACD/PBX with predictive dialing
- Dynamic Inbound/Outbound™ call management
- Single System Image View™ combining multiple sources of information into uniform user customized presentation displays.

Our desire is to provide information about predictive dialer technology that goes beyond the hardware and software that comprise the system. The people factor is an integral and critical part of the entire equation. Our core philosophy that supports transformation of a call center into a customer care center is the primary driving force behind the development of all of our products and services.

In the future, as consumers become more inundated with information it will be even more imperative for corporations to establish environments to favorably respond to the customer and treat the "Customer as King". Our goal in this book is to offer information, provoke questions and ultimately contribute and support the drive for all corporations to transform their call centers into customer care organizations.

Chapter 1

An Overview of
Predictive Dialers

THE ORIGINS OF PREDICTIVE DIALING

As some of us struggled through the late '70s trying to build an Automated Call Distribution (ACD) telephone system, a very progressive and people-caring utility company, the Wisconsin Public Service Corporation (WPSC), had a more pressing problem to overcome.

THE NEED

The area served by WPSC was affected by severe winter storms that would down power lines, blacking out thousands of WPSC's customers. This same weather made quickly restoring power to affected customers crucial.

The problem WPSC faced: how to quickly reach and dispatch maintenance personnel necessary to restore electricity. WPSC's Emergency Dispatch Center (EDC), in accordance with union regulations, had to follow a complex dialing process to locate and dispatch maintenance personnel to the affected areas. First, EDC employees called additional EDC employees at their homes and asked them to report to work as soon as possible. Once adequate personnel arrived, they would begin the manual and laborious task of calling and dispatching maintenance personnel. Complicated union regulations required maintenance personnel be contacted in reverse order of seniority. This meant high seniority maintenance personnel could be contacted only if after all lower seniority personnel were called, not enough maintenance personnel were available. To make matters worse, some maintenance personnel would try to avoid coming to work by not answering their phones.

Meanwhile, as EDC personnel were making thousands of calls to locate and dispatch maintenance personnel, WPSC customers were still waiting, and freezing, in their increasingly cold, blacked-out homes.

THE SOLUTION

Wisconsin Public Service Corporation committed its resources to develop a solution to quickly reach and dispatch maintenance personnel to restore affected customers' power. The system they designed not only solved their immediate business problem. It spawned the technology this book begins to explore.

WPSC's system integrated an automated telephone call processing system with an IBM 3093 mainframe database. With only one agent, the system, called CompuDialer, was capable of delivering telephone calls to maintenance personnel 24 hours a day, 365 days a year. This substantially lowered EDC operating costs.

CompuDialer worked the same way EDC personnel would, except it made calls more quickly with less personnel and risk. It would dial telephone numbers simultaneously on four lines, listen for a human voice to answer (while discriminating answering machines and other telephone line signals), then connect the EDC agent.

The EDC agent's terminal would display the basic information of who was connected, the status of the other three lines and the calling list disposition. Sometimes, the system would keep additional maintenance parties on hold until sufficient staff could be assembled and dispatched to affected areas.

The system had to make some basic predictions: how many maintenance people would be at home to answer calls, and how many calls to launch to get sufficient personnel to the affected areas as quickly as possible.

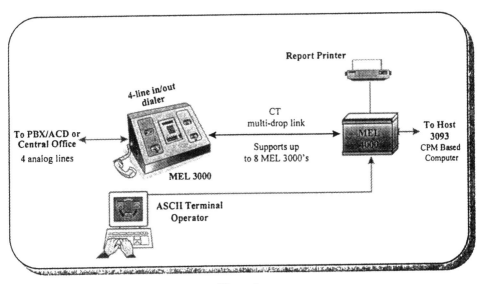

Figure 1

Year 1980, World's First Multi-Line Automated Dialing

Computer Telephony System

THE BASICS OF DIALING

As mentioned above, the basic automated dialer technology was created in the late '70s with the main purpose of locating people through the use of multiple telephone lines. Automatic dialers could do the work of many people, in less time, for lower costs. They could complete the people-intensive task of calling hundreds, even thousands of people, utilizing an agent's time only when a party answered the phone.

HOW AN OUTBOUND TELEPHONE CALL IS MADE

To reach a person on the telephone, we often ignore that we must go through many steps. Examining a typical after-hours call to an at-home colleague, let's look at what's involved:

STEP

1. Locate the nearest phone
2. Look up their home telephone number
3. Pick up the handset
4. Wait and listen for dial tone (if no dial tone, hang-up and try again)
5. Dial each digit of the telephone number
6. Wait and listen for a live voice answering the call (if no signal or voice is heard, i.e. line is dead, hang-up and try from step 3; if a busy signal is heard, hang-up and try from step 3, but wait a few minutes; if ringing is heard but no one answers, listen through four, five, or six rings, then hang up and later try from step 2 again; if telephone company announcement is heard (wrong number, number disconnected or number changed), hang up and try from step 2 at a future time; if after a few rings, an answering machine answers, either hang up or leave a message, or ask the person to call back at a specific time; if after a few rings, fax/modem gibberish is heard, send a fax/modem message, or hang up and try from step 2 at a future time.
7. When you hear a live voice answer, speak with the person you called, ask for that person or leave a message.

At last you actually talk to the party you intended to reach, or at least leave a message for them. If we assume you talk to the person you

STEPS	
Getting to the phone	5 sec
Looking up the number	5 sec
Listening for the dial tone	2 sec
Dialing a local number	4 sec
Waiting through 2-3 rings	15 sec
Subtotal	**31 sec**
Talking with desired person 2 minutes	120 sec
Total Time	**151 sec**

called for about two minutes, let's look at how much time you spent just processing the call — assuming the person you intend to reach answers the phone.

In the scenario you spent 31 seconds simply placing the call. In most cases, a predictive dialer can automate these steps, reducing them to 5 to 7 seconds. Since the conversation lasted only 120 seconds and getting the person took 31 seconds, approximately 25% of your time was wasted. Now multiply that 25-second time savings by the total number of calls all your agents make each week. Convert these seconds to hours, then divide that number by the 40-hour work week. This is how many agent hours are spent on nonproductive activities.

Let's take this example a little further. What about the situations where, after step 5, you hear a busy signal? And what if you have to retry the number a couple of times, each time waiting a minute or more between attempts? Look what happens:

STEPS		
1.	Get phone	5 seconds
2.	Find number	5
3.	Listen for dial tone	2
4.	Dial number	4
5.	Hear busy signal #1	2
6.	Hang up	1
7.	Wait 1 minute	60
8.	Listen for dial tone	2
9.	Redial using speed dial	1
	First busy scenario subtotal	*82 seconds*
10.	Hear busy signal #2	2 seconds
11.	Hang up	1
12.	Wait 1 minute	60
13.	Listen for dial tone	2
14.	Redial using speed dial	1
	Second busy signal subtotal	*66 seconds*
15.	Ringing	6
	Person answers after 1 ring	
	Total processing time	*154 seconds*
	(2 busy signals)	

In the first busy signal scenario, measure 82 seconds of nonproductive time against 120 seconds of actual productive talk time. You'll see that approximately two-thirds or 67% of productive time is wasted just trying to contact the person. In the case of redialing twice, with 154 seconds of nonproductive time, well over 100% of productive time is wasted. A company using a predictive dialer could talk to at least two people in the same time it took the human agent to reach one person.

AUTOMATION OF CALL PROCESSING

Applying the manual dialing process to business applications, where hundreds of thousands of outgoing calls are made every day, one can easily see the enormity of time spent on nonproductive call processing. If only the manual dialing steps were automated, alleviating much of the nonproductive time, from 20 to 100% of this wasted time could be reallocated toward quality talk time with a customer. When one considers a business-to-consumer call where the agent must first select the consumer's record, then review that record (typically such prereview process takes 30-45 seconds), then dial the telephone number (i.e. go to step 3 in above example), the automation of call processing yields even greater benefits. Typically a predictive dialer creates from 150 to 400% efficiency gains.

WHY AUTOMATED DIALING

Most of us experience the benefits of automated dialing (without thinking about it) every time we use the speed dial or redial features on our phones. On a larger scale, the objective of most automated dialing equipment is to relieve the user of tasks that can be handled either unattended or more efficiently by a computer. This makes the time spent on the phone more productive.

As a standalone solution, the predictive dialer offers three powerful advantages over manual dialing:

1. With the exception of interactive conversation, a predictive dialer can manage all the tasks associated with making a call.

2. The predictive dialer can also queue new calls by using statistical averages to predict when an agent will complete the current call. This technology enables the dialer to begin dialing new call attempts before the current call is finished.

3. Predictive dialers increase efficiency by simultaneously delivering the voice call to a telephone (or headset), and that consumer's record to a data terminal. This increases the average productive talk time per agent, per hour, to well over 50 minutes compared to an average 10 to 15 minutes in a manual environment.

NEED FOR CALL PROGRESS DETECTION AND VOICE MESSAGING

Automating the calling process requires the ability of computers to recognize such phone signals as busy, ringing with no answer, various SIT tones provided by phone companies, and an actual human voice answering. Various techniques are used to monitor call progress on the phone line. One of the earlier techniques developed, Cadence Signal Detection (CSD), still in use today, has been deployed successfully in over 50 countries worldwide.

Allowing the computer to "learn" signals (such as busy and ringing) being heard on the telephone line, and storing these sounds as templates, became an important discovery in the late '70s. Comparing the results of stored templates of sounds on an ongoing basis, from one time frame to another, enables the algorithm to detect variations in signals. Through the Cadence Signal Detection algorithm, various tone and voice patterns can be detected. The discovery and implementation of Cadence Signal Detection enabled the automation of call processing.

This same digital technology that is capable of monitoring, detecting and storing sound templates, is also used to deliver messages as needed. During the process of automatic dialing/predictive dialing, situations occur where a recorded announcement needs to be delivered to the called party. Because the voice messaging technology built on Digital Signal Processing (DSP) technology is the same as the CSD — the system can simultaneously deploy both technologies during a call. The user of the predictive dialing system can simply specify the country and language of choice for message delivery. The system automatically does the rest.

NEED FOR CALLING LIST PREPARATION AND MANAGEMENT

For a typical business application, customer and prospect information is normally kept in a database. Businesses generally have mainframe computers with legacy databases or newer relational databases such as Oracle, Sybase or DB2.

Contacting existing customers requires extracting a list of customer records. The application software responsible for record management typically contains user definable selection criteria, allowing extraction of a group of customers ideally suited for a specific campaign. A company can also use that selection criteria to mark existing records that should be included in a predictive dialing campaign.

Lists and record management play an important role in contacting prospects and customers. This is why there are many tools, software applications and outside consultants to assist companies in keeping their customer and prospect information up to date.

An important part of calling list management is the ability to work with a multitude of databases and host applications. Information on prospects and customers is often kept in multiple external information systems. Calling list management must access all these information sources, combining them effectively to create a final list, or lists, used for outbound dialing.

In financial and customer services applications, our experience shows that once lists are transferred to the dialer, there are additional needs that require further selection and sorting to successfully complete even more focused campaigns. Often this list management activity must take place in real time. Usually, the predictive dialing server allows the call center manager to examine the entire list of records, then select the actual records that are going to be dialed. This process is completed while the system is dialing and other call-processing activities are taking place.

NEED FOR MANAGING MULTIPLE TELEPHONE LINES AND AGENTS

Unlike the Wisconsin application that used a single agent, most predictive dialer applications require the management of multiple agents, positions and telephone lines. Imagine trying to reach and speak to 2,000, 5,000 or 50,000 people in one day! A single agent could not do it.

Building a switch (like a PBX) capable of connecting the telephone line on which the dialed number was answered, to the next available agent — all in a fraction of a second — had to be undertaken. This single issue of virtually instant connection created the demand for a predictive dialing switching platform. Prior to this, no PBX/ACD was capable of making the fast connections required by a predictive dialing application.

Certain intelligence and logistics are necessary within the call management portion of the predictive dialing system. The call management subsystem needs to know which telephone lines are used for dialing local numbers, which for long-distance calls and which for international calls. Further, some telephone lines are managed on a global polling basis, while others can be dedicated to allow predictive dialers to place outbound calls and/or receive inbound calls.

In addition to managing the phone lines over which a campaign is mounted, predictive dialers need to manage the agent positions. To accomplish this, agent positions must be dynamic. A predictive dialer needs to associate a specific outbound campaign or set of campaigns with specific agent positions. Agent positions need to not only process outgoing calls, but receive incoming calls from both inside and outside the call center. Finally, the system needs to provide floating workstation management. Virtually any properly equipped workstation, interconnected through a Local Area Network (LAN) or Wide Area Network (WAN), should be capable of accessing predictive dialing resources on demand.

NEED FOR PREDICTABILITY

By automating the process that connects an agent to a targeted party, predictive dialers can greatly increase agent efficiency. Predictive dialing's real challenge and promise, however, is taking the process one step further to ensure that the moment an agent completes a call, they are connected with a new party — keeping each agent as close as possible to 100% productive.

To accomplish this, predictive dialers must predict the exact moment an agent will complete a conversation (or transaction). With this time in mind, the system begins placing enough outbound telephone calls on one or multiple telephone lines so the moment the agent becomes available to take the next call, a targeted party is just answering the phone. Virtually perfect predictability eliminates the unproductive time agents spend between calls.

Since agents typically have their own pace time and differ in the way they handle inbound/outbound calls, the system must predict individual agent availability to take such calls. Time of day, day of week and type of campaign all make a difference. Predictive dialers must track and apply all this information to individual agent characteristics.

NEED FOR INFORMATION ACCESS

Even more important than getting lists downloaded and segregated into campaigns, is making sure once the prospect or customer is reached, information on them is immediately accessible. This requires the agent's workstation platform to access different types of information from different sources. We call this Universal Access™ and/or Single System Image View™.

Additionally, the agent's desktop must guide them through the campaign, which can be quite complex. And it needs to filter information from different media sources. With the information super highway, Internet and all future information mania we are sure to experience, it is absolutely critical that the agent workstation be an intelligent PC-like device running application programs, and not an outdated ASCII terminal.

Today's desktop PCs are equipped with multiple-host accessibility, multimedia capabilities and intelligent workstation capabilities to help agents filter through the information they are provided to make better and faster decisions. The need for information management and Universal Access is paramount and should be a key strategic requirement in putting together a predictive dialing solution.

PUTTING THE PIECES TOGETHER

Now that we've identified some of the key needs, let's look at how predictive dialing actually works. To do this, we need to look at its building blocks or system architecture.

SYSTEM ARCHITECTURE

The predictive dialing platform is created by combining various technologies, the call progress detection piece, telephone lists and records management, the switching platform and the workstations. Its underlying architecture is the key to the predictive dialer's expandability, remoteness, flexibility and application interoperability across an enterprise.

Predictive dialers need to be constructed using a building block approach. They must be built as open systems, with client/server distributed processing. This enables various components to be enhanced and updated as new technology becomes available. Which in turn gives users the benefits of incremental technology advances, rather than making them wait for enough revolutionary changes to warrant replacing the entire system.

Predictive dialers employing building block open architectures can also be made up of hardware and software platforms best suited for the tasks required of a given component. The final result is a modular system that meets industry standards while incorporating individualized components that can grow in size and application scope.

COMPONENTS OF A PREDICTIVE DIALER

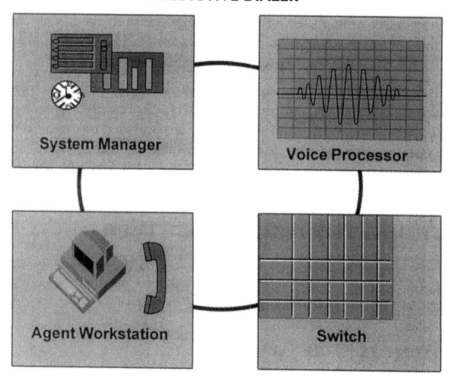

Figure 2

System Architecture

The major components of a predictive dialing system are:

SYSTEM MANAGER

It monitors and controls the enterprise-wide applications where calling campaigns are generated. The commands are issued to voice processors/switching platforms and CTI servers. This is also where pacing, Strategy-Flow™ rules are set and customized on-line reports are provided.

VOICE PROCESSOR

It handles all interactive voice processing applications, automatically processing and monitoring the progress of each call, once dial-out commands are received from the system manager. It distinguishes between call-processing signals, answering machines, modems/faxes, SIT tones and live voice answers. It intelligently transfers calls throughout the enterprise and provides interactive voice messaging/response applications.

SWITCH

It routes voice (and optionally related data) through intelligent switching. With instructions from the system manager or telephony server, it sends connected outbound and inbound callers to appropriate agent positions across an enterprise.

AGENT WORKSTATION

It deals with complex data/multimedia interactions between the agent and the multitude of user applications across an enterprise.

SYSTEM CAPACITY AND INTEGRATION

A typical system needs to accommodate call centers that may begin with a few agents and grow to several hundred. It may start by networking several predictive dialing systems together within one location, then expand to include multiple remote locations. Through modular building block design and open systems architecture, growth can be as economical as it is unrestricted.

PREDICTIVE DIALING — HOW IT WORKS

The least productive part of making a large volume of calls is dealing with those that result in a busy signal, bad number or no answer. Predictive dialing systems recognize various signals, intercept tones returned by the telephone company and automatically process them. Current call process monitoring techniques accurately and capably distinguish between live voice, answering machines and other signals.

Through audio-signal or voice-energy pattern analysis, predictive dialing systems not only dispose of no-contact call attempts but, depending on the initial result, schedule numbers to be recalled. For example, call attempts resulting in a busy signal may be scheduled for immediate recall, while call attempts resulting in no answer are rescheduled for thirty minutes or an hour later. The system tracks and reports on the result of each call attempt, so users can manage and update their calling strategy.

Because a percentage of calls are invariably unsuccessful (result in no contact), the predictive dialer simultaneously initiates call attempts on multiple lines, anticipating that some attempts will not get through. The number of simultaneous calls initiated depends on the success rate (or hit rate) at that moment. For example, when an average of only one of two calls results in a contact (50% hit rate), the system generates two call attempts for every agent.

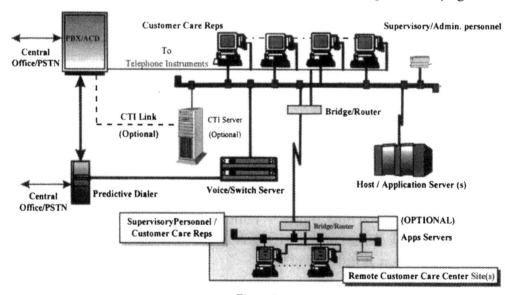

Figure 3
Typical Call Center Environment

This concept is similar to the way airlines overbook flights, knowing that historically there is always a percentage of no-shows. The science of overbooking makes sure there are not too many passengers and not enough seats.

This is one area where predictive dialing vendors establish product differentiation. The key is maximizing productivity while minimizing nuisance calls. (A nuisance call is when a contact is established, but no agent is available, so the system must terminate the call.)

There are options for eliminating nuisance calls even when users push the system for maximum talk time. Alternatives to hanging up on the called party include playing a recorded message, putting the called party on hold until an agent is available, sending the call to an overflow or standby agent or actually canceling the call after it has been dialed. Using statistics and timing, telephone calls in progress can be terminated before a person answers the call.

Finding the right balance between agent productivity and nuisance call generation is an ever-changing business decision call center managers make daily. The best predictive dialers help call center managers maximize productivity, while minimizing nuisance call generation.

Restrictions imposed by regulatory agencies regarding playing recorded messages, require that telemarketing applications generate very few nuisance calls. Most business managers are willing to accept some nuisance calls to achieve maximum productivity. But there are options (usually at a slightly higher cost) for those who require that no nuisance calls be generated.

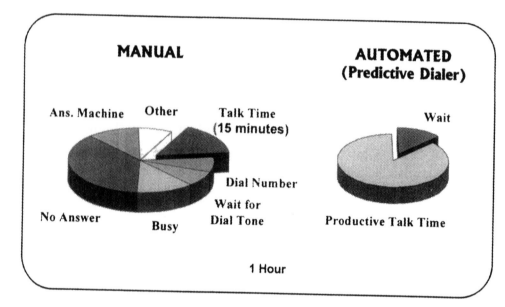

Figure 4

Manual vs. Automated Dialing

Another area of emphasis is at the workstation, where multisource access, full scripting, dynamic branching and data capture are just some of the key features required. Currently, most data-entry tasks are automated at the agent's workstation through function keys on the data terminal or custom application software running on PC workstations.

Systems can be created that automatically access multiple databases to provide agents with necessary information on a particular party. Systems can also update information stored in multiple hosts, saving the agent time and decreasing the chance of errors during data entry.

Removing the remedial tasks of manual dialing lets agents spend more time doing what they've been trained to do — talk with customers. This inevitably results in a boost in agent morale.

SEAMLESS INTEGRATION

Functionality and features vary between predictive dialer vendors. Typically, predictive dialing products provide real-time and historical-management information, plus the flexibility of allowing the user to control and monitor system activities. Beyond user features is the ability for the business or call center to integrate and work seamlessly. Here the difference in predictive dialing systems can be dramatic. Underlying architecture must follow open systems standards. This allows access to diverse information sources over local and wide area networks, seamless integration with legacy mainframes, enterprise-wide contact management, teleservicing and order-entry-type applications.

The benefits of predictive dialing include:
1. Protection of existing and future investments
2. Faster access to information
3. Improved customer service that competitively differentiates a company

The predictive dialer is a critical piece of an integrated enterprise-wide solution — one that helps get the job done faster, with fewer people, at a lower cost.

THE BUSINESS NEED

Today's businesses have more choices in how they communicate with their customers. Businesses can use mail, E-mail, fax, radio, television, telephone, video tapes, audio tapes, compact discs, books, magazines, the Internet, satellites and the list keeps growing. All are geared toward a single purpose — Message Delivery!

Many of these communication methods depend on timing and interpretation. For example, if we don't listen to the news or read the paper, chances are we will be uninformed about today's headlines unless someone tells us about them. Now consider the telephone's unique advantages. We can prepare a special campaign where certain people are chosen to receive a phone message say between 7:00 p.m. and 9:00 p.m.

Such campaigning, if automated, using multiple telephone lines and multiple agents, can directly deliver such information to every person answering the phone. The agent can even personally answer any questions. If the call recipient prefers to receive this call at another time and/or place, the agent simply schedules a call back.

The telephone is the closest and most direct communication media short of being physically next to the person. This is why the telephone has become the preferred method for achieving most business communication goals.

In the late '70s and early '80s, businesses and institutions from Sears to Wal-Mart to CitiBank/CitiCorp, realized the telephone was the best and most cost-effective method for staying in touch with their customers.

Figure 5

Business Communications

Automatic Call Distribution (ACD) inbound call processing was created at this time. Customers and prospects received 800 numbers to call for service orders, changes and reservations. Call centers began processing thousands, even hundreds of thousands of inbound calls a day. Though ACD innovation and development is not the subject of this book, we must give credit to the many creative and talented people who contributed to this telephone call processing revolution.

Look closer at inbound (ACD) call processing applications such as service and orders. These are more reactive type applications, where the customer and/or prospect takes the initiative to call the business.

Developing lasting relationships with customers must also be proactive. Businesses must use the phone to inform customers and prospects, obtain feedback and promote products and services on an ongoing basis. The result: the proliferation of outbound predictive dialing solutions into virtually every business sector.

The two most recognized sectors that clearly justified the early deployment of predictive dialing were and still are:

- Financial
- Telemarketing/Telesales

We will take a brief look at both of these in explaining the need for predictive dialing solutions.

FINANCIAL SECTOR

Banks, mortgage companies, credit card issuers and auto-finance companies are just some of the financial institutions with large debt problems looking for a solution. Gradually the delinquency rates on credit accounts have risen alarmingly, approaching 10% of the overall customer population base for some card issuers. If we look at an institution carrying and processing say a million credit cards a year, even a 5% delinquency rate translates to 50,000 accounts making no payments or payments over 120 days late.

Without consistent follow-ups and reminders, many of these accounts soon move into the prelegal or write-off stage. The amount of money delinquent and outstanding in the USA is astonishing. In 1993, it totaled nearly $80 billion dollars. With an average of $2,000 in late payments, just the 50,000 above delinquent accounts translates to $100 million in delinquencies.

SO HOW DOES THE PREDICTIVE DIALER HELP IN COLLECTION EFFORTS?

Many financial institutions develop a front-end collection process.

They create and maintain large customer databases, monitoring and tracking the habits of delinquent account clients. Many advanced behavioral modeling applications have surfaced, allowing financial institutions to score their client base to better predict who may be a candidate for late payments. Similar methods are used during the application procedure, where prospective clients are evaluated for their ability to pay on time.

But no matter how the client selection process is done, some clients still pay late, or choose not to pay at all. We must recognize that most people carry multiple credit cards and many have multiple loans. Getting to these customers ahead of other creditors is critical. Tactfully reminding these customers early in the delinquency stage about how much they owe and when their payment is due is also critical. Sending a reminder notice in the mail is not very personal and is rarely effective.

SOLUTION

The best and most effective way to reach the people with chronically delinquent accounts is by telephone. Why? Well, the telephone is personal. It lets an agent speak directly with the person so the agent can explore various tactics.

For example, the agent can discuss mutually acceptable alternatives such as smaller payments over a longer period. (Yes, people talking to each other, more often than not, can compromise.) Perhaps the payment is due on the 5th of each month. The agent could change the due date to the 1st or 15th if either of these work better. Often by just getting to the person first (before other creditors), a level of commitment is established that provides a higher likelihood of payment.

Also, by calling on a specific day, sometimes even at a specific time and place, the agent can regularly confirm that payment has been sent. This can serve as a constant reminder and reinforcer.

COSTS

Mail and telephone calls are typically used to solicit payments. To reach customers in a timely manner, usually several telephone call attempts must be made. Sometimes, the agent may need to call the person at home and at work over a period of several days. During this time, many unsuccessful attempts may take place. The number could be busy and later, answered by a machine. The agent may need to leave a message, trying to schedule a specific time and place to speak with the person.

The process of getting to a specific person in the shortest time possible could translate into high costs. First, the many call attempts imply higher telephone/long distance costs. Second, the increased number of calls and their associated scheduling suggest an increase in personnel.

By deploying a predictive dialing solution, multiple call attempts can be made to every delinquent account holder as often as desired. A predictive dialer plans the work, dials and redials the telephone numbers, searches for the person by calling multiple numbers and finally gets the desired person on the line. Best of all, the predictive dialer accomplishes all this without human assistance! In fact, until someone actually answers the phone, the operational costs of reaching delinquent account clients are virtually eliminated.

Let's look at our previous example of 50,000 delinquent accounts assuming the following scenario:

1. Must reach every person twice a month
2. Average talk/connect time with the right party contact = two minutes
3. Average connect time with a wrong party contact = half a minute
4. Operational hours/month = 172 hours (6 days x 7 hr. x 4.3)
5. 50% of contacts are with the right party, 50% with a wrong party
6. All answering machines, busy signals, no answers, telco messages and modems/faxes are screened entirely by the predictive dialer (agents are not connected to such calls)

With the requirements of reaching 50,000 customers twice a month, 100,000 direct conversations with a right party must take place each month. Also because of the 50/50% split statistics, 100,000 conversations with a wrong party will take place. Looking at the typical 25% efficiency of manual dialing, this translates into 1/4 hour or 15 minutes of productive time per agent hour (i.e. time spent by an agent talking to a right or wrong party).

Since each agent position works 172 monthly hours, with 1/4 of this time spent talking with the customer, only 43 hours or 1,290 right party contacts a month per agent can be achieved. To process 100,000 contacts would require approximately 77 agent positions, just for handling right party contacts. Similarly, to speak with 100,000 wrong party contacts would require approximately 19 additional agents, totaling 96 agents dialing manually.

Using the predictive dialing solution, however, with agent efficiencies in the 85% range, only 29 agent positions are required. These positions could further be reduced through intelligence and automation software at the workstation. (See Chapter 4 - "Hypothetical Company" for calculation.)

To reduce agent phone transaction processing time, developers have streamlined software, eliminating many of the steps previously required during a call. Solutions use advanced Human Interfacing (HI), Graphical User Interfacing (GUI) and automated resource management to reduce application complexity and gain an additional 10 to 50% in agent desktop productivity. (See Chapter 7 - "Single System Image View.")

BENEFITS

Bottom line benefits for predictive dialing in debt collection are:

1. Higher level of customer care — keeping in touch more often and at the best time
2. Faster collection cycle — obtaining payments more quickly from delinquent customers
3. Lower operating costs — reducing the number of agents needed to do the job

Automating the debt collection process (accounts receivable) through deployment of predictive dialing can be beneficial for both business and debtor.

As we see in the previous example, fewer people are required to perform this laborious and repetitive task. We also find that many agents previously engaged in the manual collection process, become more involved and more self-confident by focusing their energy on the customer, instead of trying to reach the customer. The company benefits through happier employees, lower operational costs and more focused and timely customer care.

The customer benefits through more consistent and personal communication with the company. This gives the customer an opportunity to work out a manageable payment plan, possibly without damaging their credit rating.

Additional predictive dialing applications have been successfully deployed within the financial sector. These include:

- Customer prospecting
- Ongoing customer services — like confirmations
- Customer satisfaction follow-ups/surveys
- Product or service offerings
- Help desk support services
- Other information dissemination
- Account fraud detection

RETURN ON INVESTMENT

AS you can see, the return on investment (ROI) could be outstanding since your investment in the equipment for an agent is substantially lower than the salary of an agent.

TELESALES/TELEMARKETING SECTOR

As successful as it is in the financial sector, predictive dialing uses an entirely different application approach with equal success in telemarketing/telesales.

Historically, telemarketing has been predominately applied towards selling products or services to noncustomers. Many of us have received a late afternoon call from a company we never heard of. More often than not, the person on the other end who's trying to sell us something knows almost nothing about us. Worse, sometimes an automatic dialer calls and plays a recorded message. What is your reaction to this type call? Are you ready to stop whatever you're doing and listen?

A large portion of sales cost goes to calling on prospects. So it's not surprising that many telemarketing centers have surfaced in the last five years. Many of these businesses deploy predictive dialing solutions to increase efficiencies.

Because it is much less expensive to sell and follow-up by phone, many businesses develop their own teleservicing facilities or outsource these tasks to third party service providers. The teleservicing business is booming and more dramatic changes are expected during the next few years. But what is so different about teleservicing automation?

Since most calls are to noncustomers, information about them can be limited. Yet, the opportunity to generate new customers makes it necessary to place the call. The agent must be prepared for just about anything. For example, if the prospect says to never call him again, this request must be honored and properly handled by the predictive dialing solution.

ONLY ONE CHANCE

In teleservicing, typically 60 to 90% of solicitation calls are hung-up on. Many call attempts must be made before one successful closure takes place. These short calls translate into many additional agent positions. The specific words spoken by the agent at the very moment the prospect answers the call become critical — because this call could be the only contact the agent has with that prospect.

The need for on-screen scripting, guiding the agent from the conversation's introduction portion through positive conclusion is significant. Though it is not the intention of this book to go into the details of on-line and real-time interactive scripting (such a topic easily warrants a separate book), a few key features are worth mentioning.

DYNAMIC SCRIPTING

Depending on the input from the prospect's data record, the agent or other sources, different scripting appears on screen with the prospects name and other significant information embedded as part of that script.

USER-DEFINABLE SCRIPTS

Agents receive scripts, graphics and input/output information sources, along with logistics for supporting interactive- and multimedia-based selling processes.

MULTIPLE APPLICATION/MULTIPLE PRODUCT SUPPORT

On a call-by-call basis, the agent may be presented with a different set of products and a corresponding set of supportive screens.

UNIVERSAL ACCESSIBILITY

Scripting software needs to be self-empowered, capable of accessing multiple data sources, presenting information to the agent in the form, content and style best suited for each call. (More on the above features plus others in Chapters 3 and 7.)

MAKING THE SALE

On-the-spot addressing of a prospect's questions is the key towards building trust and comfort between agent and prospect. Crucial to providing such a high level of responsive teleservicing, is ongoing agent training. Sadly, it is not uncommon for a teleservicing campaign to run for only a few days with dozens, even hundreds of products or services being sold. Since separate training for such short campaigns may not be affordable or even practical, another solution must be found.

Through the deployment of client/server application technology, self-training while selling is possible. First, by providing desktop/workstation presentation screens that are common to all selling applications. These offer a standard set of icons/images with predefined actions. By adding interactive scripting with on-line help for products and services being sold, the agent is guided through all steps of the negotiation/selling process with little or no need for training. With this Single System Image View, agents become self-empowered and knowledgeable from the word go.

From the call processing standpoint — getting a prospective buyer to stay on the phone requires a very different call-handling process than may be acceptable when calling a delinquent account customer. For example, many predictive dialing solutions put called parties on hold when a situation arises where no agent is available to handle such an outbound call. Though in some situations, customers may not mind occasionally being put on hold by their service providers, putting prospective customers on hold would be totally unacceptable and, in most cases, irresponsible.

Can you imagine answering a call at home and hearing a recorded announcement, "This is Company X, please hold"? People typically hang up on such calls. But for the few calls that are accepted, imagine the problems for the agent who is finally connected to this prospect. How will they explain to the by now highly defensive prospect why they were put on hold?

The bottom-line key benefits for predictive dialing in telesales are:

1. Ability to reach a large group of targeted prospects in a fraction of the time needed for manual dialing
2. Capacity to increase sales revenues over a shorter time frame
3. Lower operating costs per sale as a result of reducing the number of agents or keeping the same staff and increasing the number of sales

CHAPTER SUMMARY

No matter what business you're in, if you offer products or services to customers, you can successfully deploy a predictive dialing solution. Businesses all over the world are transforming themselves into customer care organizations, taking care of customers every step of the way.

All your employees become, in essence, Customer Care representatives, with responsibilities and authority to maintain and revitalize relationships with customers for life. Predictive dialing will help them do this much more efficiently and economically. This will pay big dividends in customer retention.

Chapter 2

Why and Where To Use Predictive Dialers

Some years ago, an associate and I got into a taxi. The cabby said, "Where to?" Quickly my associate replied, "Take us anywhere. We've got opportunities everywhere." This simple quip accurately describes the opportunities available to those companies that use predictive dialing.

MORE EFFECTIVE THAN MAIL

First let's examine why it's more effective to communicate with customers and prospects using the telephone as opposed to sending mail:

- Phone calls cost less than letters.
- Phone calls are faster than letters.
- A letter may generate an inbound phone call. Now we have the cost of the letter plus the cost of a phone call. (Remember the major cost of a phone call is the agent's time. If 800 numbers are used, add to the cost of the call itself, the on-hold time spent while the customer waits for an agent to answer.)

- A letter may generate a letter back. This is even more expensive to process than the original outbound letter and may create a need to send another letter or an outbound phone call. All of this processing is people intensive.

- With a phone call, you get instant feedback. But just making the phone call doesn't mean automatic success. You might find that:

- The phone has been disconnected.
- The customer is no longer employed where he/she last reported.
- The customer no longer lives at the address in your database.
- The customer has fallen into serious financial difficulties.
- The customer is seriously ill.
- The customer has died.

Most companies find that information gained through these calls allows their early intervention to protect company interests. Couple this with the cost savings over sending mail, and it becomes no contest. Telephone contact is better than letters that are not legally necessary to protect a businesses' rights.

THE BUSINESS LIFE CYCLE

Looking at virtually every business with customers, products and ongoing services, it becomes very clear, very fast that customer contact is crucial to every major step in the business process. Examine this process through the eyes of predictive dialing. The conclusion: successful business life without ongoing customer contact cannot exist.

Figure 6
Looking at the Business Lifecycle

Let's briefly look at some of the key business life cycles directly affected by predictive dialing solutions:

- For a business to grow and attract new customers and prospects, outbound calling campaigns need to be deployed.

- Once customers are selected and prequalified, telesales/telemarketing campaigns can be launched to promote and sell products or services.

- Once products or services are sold, properly timed customer follow-up campaigns can be launched to assure the highest level of customer satisfaction.

- As products or services are deployed throughout the customer base, predictive dialing manages inbound and outbound customer support services, assuring prompt response and satisfactory resolution to any issues.

- Having satisfied the customer base, new telesales/telemarketing campaigns can be launched, selling value-added products and services.

- In cases where customers become delinquent on their payments, reminder campaigns can be deployed early on (3 to 7 days past the due date), promoting customer care and continued business relationships.

- Once customer accounts reach 30-45 days overdue, debt collection campaigns can be deployed.

- On a proactive basis, numerous customer awareness campaigns can be occasionally launched, educating and informing customers of policies, procedures and significant events.

- On a proactive basis, customer surveys can be deployed (both inbound and outbound), giving customers an opportunity to share views on products/services and their vision for the future.

This circular-like business process continues as long as the business thrives on improving customer communications. The business life cycle never ends. And especially for growing businesses, this multistage cycle repeats itself sometimes several times each year.

Predictive dialers can be used by virtually any business. Let's look closer at some real business opportunities.

MAKING BUSINESS-TO-CONSUMER CONTACTS

The previous chapter dissected the dynamics of outbound calling with and without the use of a predictive dialer. In general terms, a predictive dialer in business-to-consumer contact increases agent productivity 150 to 400%, depending on the characteristics of the calling application. In other words, if 60 agents are now making outbound business-to-consumer calls, a 150% improvement requires just 24 people to accomplish the same work; at 400% improvement, only 12 people are needed. Therefore, in six to twelve months, personnel savings alone generally provide complete cost recovery of a predictive dialer.

Here's another way to look at predictive dialers for companies unable to make all their required contacts with their current staff:

- If a business currently makes 4,000 contacts per month with existing staff, a 150% improvement would increase that to 10,000 contacts per month with the same staff.

- With a 400% improvement, those 4,000 contacts become 20,000.

When we talk about business-to-consumer contact, we're talking about reaching the average John Q. Public, most likely at home. However, reaching this typical consumer poses some problems. When will that person be home? Is it necessary that we speak with that specific person, or will a spouse or any adult who answers the phone be satisfactory? Is it acceptable to leave a message on an answering machine? The answers to these questions depend on the calling mission's purpose.

Let's briefly discuss some typical applications of predictive dialers and see how they differ.

USING PREDICTIVE DIALERS IN COLLECTIONS

Many rules are associated with making collection calls to delinquent account customers. If the call relates to a past due payment, the law and/or collections industry associations require that the matter be discussed only with the person responsible for the debt.

For example, on a past due credit card payment, we must speak with the person to whom the card was issued. This is true even if additional cards were issued on the same account to other family members. If, however, a cosigner was involved, the law allows us to speak with them. It is entirely possible that the cosigner has a different home and, therefore, different contact phone number.

In mortgage loans, usually a spouse or significant other is required to be a guarantor on the loan. In this case, it is perfectly acceptable to speak with either party about a past due payment. This policy also frequently applies to automobile and certain types of personal loans.

In essence, all collections contacts are considered a private matter. It is a violation of privacy laws or industry practices to discuss financial matters with anyone other than the principal(s). On the other hand, leaving a message for the responsible person to return a phone call is perfectly acceptable as long as the nature of the call is not divulged. Acceptance of this tactic by the industry created the need for predictive dialers to accept inbound calls.

SHORT-TERM OVERDUE

A typical outbound call scenario might go like this:

(Phone Answerer)
"Hello."
(Phone Caller)
"Hello, this is Maxwell Smart from Best Bank. May I please speak with John Doe?"
(Answerer)
"No, I am sorry he isn't here now."
(Caller)
"Well, could you please get a message to Mr. Doe to call the Best Bank today at 800-555-4321?"
(Answerer)
"I'm not sure. I might."
(Caller)
"Thank you. Just in case Mr. Doe can't return the call, could you please give me a time today that I might be able to reach him at this number?"
(Answerer)
"Yes, he should be here by 6:30 p.m."
(Caller)
"Thank you. If he is unable to return my call, we will try him at 6:30 p.m. Good-bye."

Obviously, this dialogue has created two opportunities for contact. First, John Doe may return the call; and second, we have a time when John Doe should be available. The key is to know if John Doe returns the call, so we know whether to call again at 6:30 p.m.

If Mr. Doe does call back, then the inbound call is routed to Best Bank's predictive dialer so it knows that the scheduled 6:30 p.m. outbound call can be canceled. This technique is gaining wide acceptance, especially at the earlier stages of collections (generally 60 days or less). Most collections departments are reluctant to contact people at a work number for early delinquencies.

First, in today's world of meetings, voice mail and travel, the odds of actually reaching a targeted person are not good. Second, many workers are unable to take personal calls at work. Third, at the early stages of a past due bill, customers may take serious offense at a call to a work number. We must remember that all Best Bank wants is a payment; they don't want to lose a customer! This is especially sensitive if there is a likelihood that customer has other accounts in perfectly good standing. And finally, it may be difficult to leave a message at work for John Doe without either divulging the nature of the call, or being too evasive. In either case, the call might endanger John Doe's employment or his privacy rights. This could create gigantic legal problems for Best Bank.

One of the most innovative uses of predictive dialing in collections is the early delinquency calling technique. With this procedure, the bank uses a pseudo collection scoring system that selects moderate and higher risk accounts that are 10 days past due. The agents first call the home. If the agent reaches the right person, he/she urges the customer to pay within 10 days. If anyone else, or an answering machine answers, a simple message "Please tell Mr. Doe that Best Bank called." is left. The customer is not asked to call back.

This reminder call, meant to be a memory jogger, also alerts John Doe that maybe in the future he better pay Best Bank on time. Banks that use this technique often notice a decrease in the float (measured in days past due) by getting those who overuse the grace period, to pay earlier.

LONG-TERM OVERDUE CUSTOMERS

There is a distinct difference of methodology on the effectiveness of predictive dialing on longer-term collections. For accounts 120 days past due, there seems unanimity that predictive dialing (and use of the telephone as the primary contact method) is not adequate.

At these late stages of collections, experts believe a single collection specialist is needed to manage the account. Drastic action is usually required: legal notification, maybe personal visits, coordination with local authorities, investigators, or even reclamation of assets. This is clearly not a telephone-contact-only application.

It is the period between 60 and 120 days that the divergence of opinion on the impact of predictive dialing occurs. Many collections managers feel that a designated specialist handling each account is more likely to result in acceptable resolution than contact by any available specialist. The key question: is the write-off reduced/minimized with this approach more often than with the contact-by-a-pool-of-collection-specialists approach?

Companies that have tried the pool-contact approach using the pre-dictive dialer tell us they notice no change in the effect on write-off. However, the productivity did dramatically increase in relation to the reduced personnel costs. Debt collections are one of the major world-wide uses of predictive dialing. The ability to react quickly to past-due payments has altered the fundamental business strategy of many lenders. With a method to handle past dues without increasing collections personnel costs, many lenders have lowered their test-marketing campaign financial requirements. If these lower requirements produce a net profit, then the predictive dialer becomes a significant competitive weapon in testing new markets/products.

USING PREDICTIVE DIALERS IN CUSTOMER SERVICE

In today's competitive world, brand loyalty is blurred. From credit cards to airlines to long- distance telephone service, the customer's universal war cry is, "Why not buy the low bidder? Aren't they all the same?" Many of these products and services actually are very close in quality. Does this mean the entire free-enterprise system must compete on price alone? Well, let's hope not!

What about service? Good service is measured in the eye of the beholder, using criteria like:

- Does the business provide 24-hour telephone access?
- Can they answer my questions and resolve my problems accurately and quickly?
- Do they keep me informed?
- Do they follow up to make sure I'm satisfied?
- Do they let me know they care and appreciate my business?
- Even though I'm just one of thousands (or millions) of their customers, do they make me feel like an individual?

How do businesses best handle these questions?

Certainly no single medium of communication will do. Communicating with the customer involves the coordinated use of inbound and outbound phone calling, letters and faxes. E-mail and Internet-based services are becoming significant vehicles.

Of course, one of the challenges will be determining which communication method is most appropriate for each customer. The appropriate vehicles may vary from customer to customer, and most certainly by business transaction, as well as subject matter of the communication. Additionally, there may be a need to deliver a message at a specific time and place.

Many progressive firms are planning their customer service future and looking at criteria needed to be a good customer service provider. Companies who deliver quality service:

- Value loyal and/or profitable customers.
- Consider every customer as a partner and potential candidate for multiple products and services.
- Offer many value-added products and services.
- Realize their products and services may not differ from their competitors.
- Understand that products and services will come and go, but having good customer relationships will assure continued sales.

- Know their customers are being offered aggressively-priced competitive products.

A concerted campaign to create customer loyalty starts with basic information on each customer and an understanding of how that data relates to the company's business. Where does this information come from?

Mailed out surveys are one source, but are rarely returned in enough quantity to be useful. Carefully constructed telephone surveys that gather information walk the fine line between inquiries on acceptable subjects and perceived invasion of privacy. On the other hand, they also often carry the message, "You're important to us, we appreciate you and your business."

PREDICTIVE DIALING CAN HELP

Using predictive dialing to gather survey information is just another way the dialer benefits a company. Consider this predictive outbound call scenario:

(Phone Answerer)
"Hello."
(Caller)
"Hello," this is Maxwell Smart, Best Bank preferred customer service specialist. May I please speak with John Doe?"
(Answerer)
"Yes, this is John Doe."
(Caller)
"Mr. Doe, I want to thank you for your confidence in us. You are one of our most valued customers." (Pause to see if Mr. Doe responds; if not continue.) "We are constantly looking for ways to provide better service to you. Could you spare me four minutes so we could find out more about how we could best provide service to you? For example, what kinds of situations should we contact you about? Which of these ways of contact do you prefer: telephone, letter or fax? What is the appropriate address and phone number at which to contact you?"

(Now Mr. Doe can say,)
"No, not interested," or "Now is not a good time; call me at home after 6:00 p.m." or "Sure, I'll be glad to"

A carefully constructed "Thank you, you're important to us." message clearly states the caller's request, tells the customer how long the call will take, explains what types of questions will be asked, and gives a business the perfect opportunity to gather personal-preference information about its customers.

This type of customer information helps businesses in endless ways. Just by knowing more about its customers, a business can increase customer awareness, improve products and services, and therefore, sell more.

From a survey, one can learn:

1. What categories of interest a customer has.
    (Sell them new products and services)
2. Which categories to send information on and, which to call on.
    (Saves costs involved in cross-selling products and service)
3. When and to where calls should be placed.
    (More efficient telemarketing)
4   And if it is, or is not okay to contact customers from time to time for additional feedback on how to best service them.

HOW ABOUT SOME OTHER IDEAS:

When it comes to customer service, a predictive dialer isn't just limited to surveys. Here are other ways companies are using predictive dialing to provide outstanding customer service:

*   Making a quality assurance call to ensure that the matter John Doe called in about yesterday was handled properly and that he is satisfied with the result.
*   Following up on a service person's visit to the customer's home. Was the service person on time? Were they courteous? Was the problem fixed?
*   Checking with the customer to see that a shipped item arrived on time and in good condition.
    (Especially important for perishable or time-critical items.)
*   Notifying a customer that an international flight has been canceled or delayed, and presenting alternative routing if appropriate.

One large mail-order catalog firm in Europe calls each new catalog recipient 10 days after the catalog's anticipated receipt. This call is referred to as a Welcome Call. The purpose: to answer questions, acquaint the recipient with the catalog and company and to personalize the relationship. Following Welcome Calls, order-volume improvement is dramatic. Exact results are privileged, but this technique works so well that the equivalent of 80 full-time agents make these calls 50 hours a week.

Customer service is a ripe area for predictive dialer use. A proper mix of telephone calls, letters and faxes appeals to the customer and creates loyalty that helps fend off competitors. Telephone contact is the most effective way to get feedback. Use it!

USING PREDICTIVE DIALERS IN TELESALES

BUSINESS TO PUBLIC

Telesales, telemarketing, telesurveys, televolunteering and telecontributing are all possible aliases for telephone solicitation. These calls, which can be inbound or outbound, often elicit a number of different emotional reactions from John Q. Public. They also offer a number of excellent opportunities for businesses.

Unfortunately, telesales has been exploited by some unscrupulous operators. This has forced state and federal agencies to examine telesales practices with regulations in mind. As usual with government regulation, they get it only partially right, tend to overdo their involvement and frequently hamper legitimate business activities.

An example: In the mid '80s, the Public Service Commissions of many states enacted laws limiting the use of the "old" autodialers that immediately played a recorded message when an outbound call was answered. The objective was to prevent the use of these dialers for telephone solicitation. One state, however, went even further. They made it illegal to use this technique for any purpose unless the firm calling had written permission from the parties to be called, that a recorded message was okay. The largest school system in this state used our autodialers to place calls in the evening to the homes of children absent from school that day. Suddenly, that school system was required to get written permission from each parent/guardian to leave a recorded message, even though a conflicting directive from the state's Department of Education required that school systems use "due diligence" in notifying of absences. This is certainly over regulation that didn't get it quite right.

Today, unscrupulous operators are threatening the legitimate business-to-consumer activities of telesales. How can a company protect itself so it can legitimately use telesales in its business? Here are a few ways:

For noncustomers — use mail, print, TV or radio ads to get the person interested enough to initiate the contact. When they do, get their permission to call them on the telephone.

For noncustomers — obtain calling lists only from firms that certify all persons on the list have given their permission for the list provider to share their names with other firms.

For customers — many firms occasionally include in a mailing an opportunity for the customers to remove their names from different types of contact lists. Taking a name off a contact list usually can be done with:

- A statement by the company that it would like permission to contact the customer from time to time about something that is of interest to that customer.
- statement (similar to the above) that permission is requested to give the customer's contact information to a different company who may have a product or service that interests that customer.

Usually only a negative ("Please take my name off your list.") response is solicited using a check box and signature line on a postal business-reply card.

Generally, businesses don't randomly call their customers about new products or services. But with detailed customer demographics (as suggested under customer service applications), they now have the information they need to select customers who might be interested.

Frivolous solicitations waste company time and money and could anger customers. Conversely, products and services germane to the customer base are sales opportunities, and present a way to further enhance the customer relationship. It's an opportunity to create brand loyalty and let the customer know someone cares and appreciates them. Of course, every contact made with a customer should be seen as an opportunity to update their demographic data.

One last thought about customer data: This is a global community. The European Economic Community is developing rules on telephone contact and privacy. Many European countries are very rigid on privacy issues. Some do not even allow the use of customer information already within a company for any purpose other than the original reason the information was furnished.

It seems unlikely that this rigidity would come to the United States. However, government regulators are at work and may get involved. Every legitimate business must ensure that their practices are up to standards that make government regulation unnecessary. And it should be ready for regulations if they are imposed. Prepare Now!

USING PREDICTIVE DIALERS FOR BUSINESS-TO-BUSINESS

Up to now, we've only considered predictive dialing from business-to-consumer. What about business-to-business predictive dialing? Well, the first difference is productivity. In business-to-business applications, agent productivity increases 30 to 60%. Using the earlier 60-agent example:

- A 30% increase requires only 46 agents to do the same job.
- A 60% increase requires only 38 agents to do the same job.
- The payback period becomes 15 to 24 months. Still not bad!

But why is this? Well, predictive dialing works most efficiently when a significant chance exists that a call placed will not be answered. In business-to-consumer calls, the response is predominantly a no-answer. What happens when a business is dialed? Someone usually answers. Or the caller gets a busy signal or voice mail.

Seventy-five percent of the productivity increases of predictive dialing come from eliminating the time wasted by agents who frequently must manually dial 3 to 4 numbers and listen to several rings before getting an answer. When calling business-to-business during business hours, the call is typically answered.

Another factor that affects agent productivity is conversation length. The shorter the average call, the more productivity provided by predictive dialing. Business-to-business calls are usually longer in duration than business-to-consumer.

So is a predictive dialer in a business-to-business call environment a good investment? Sometimes yes, sometimes no. One must first determine when and when it's not beneficial:

1. Is the application itself relevant in business-to-business contact? Here the call length is important.
2. Must a specific person be contacted, or can a contact with a voice mail or within a particular department get the job done? Specific person contact is complicated by meetings, the person perhaps already being on the phone or having someone in their office, etc.
3. Does the application lend itself to periodic contact on a semi-appointment basis? For example, an arrangement is made to call every Tuesday between 2:30 and 3:00 p.m.
4. Does the application require the telephone agent to study a significant amount of information before he/she can intelligently have an accurate conversation with the party called?

CHARACTERISTICS POSITIVE TO PREDICTIVE DIALING USE

Predictive dialing is definitely beneficial if the business-to-business call:

- Lasts an average of less than four minutes
- Doesn't require a specific contact
- Lets an agent schedule another call for a specific time
- Requires little or no preview of information by the agent before dialing

If calls meet the above four criteria, productivity improvements will result. A good exclusion criteria guideline is that long calls that require substantial prereview are definitely not candidates for predictive dialing. Otherwise, each application must be individually examined. There is no cookbook formula for success in a business-to-business environment.

The call scheduling system within predictive dialers is most useful in assuring that calls scheduled for an appointed time do happen on time. One European business uses this appointed-time-calling feature to contact their small retail customers twice a week to take their orders. When the call is made:

- The agent's screen shows what products the retailer carries.
- It may also include a special promotion or new product suggestion.
- The order is entered quickly.
- It ships the next day.

Saving sales costs, making sure their product is on the retailer's shelf (not out of stock) and quick shipment and delivery add up to the kind of service that retailers require.

ENSURING COMPANY POLICY

A subtle, often unnoticed major benefit of a predictive dialer is the management control, or operational streamlining, it provides over inbound/outbound calling missions. The dialer provides management with tools to make sure rules and policies are followed by both agents and supervisors as they make their calling decisions.

Specifically, how can a predictive dialer benefit management?

- It assures that agents follow the call center's rules and policies. For example, if the call center manager identifies a specific group of people or order of contact, these instructions are automatically followed.
- It obeys the company rule(s) that all first attempt no-answers are retried periodically during the day.
- It records every call attempt made to each person, and possibly updates the host database (due diligence/audit of record keeping).
- Its work-flow control assures accurate adherence to policy,

regardless of which agent is on the phone.

- Since they no longer need to make sure telephone agents are processing their work as prescribed, supervisors are free to concentrate on agent training and coaching.
- As required by some applications, dialer contact management always executes specific time and date (scheduled) call attempts as high priority.
- If the wrong person answers a call, some dialers offer outbound/inbound (return calls) call coordination of the leave-a-message scenario.
- If a requested return call doesn't happen, it executes a scheduled call-back time.
- Since dialers know the exact length of each call, who processed it, and the result, management automatically gets standard co prehensive reports on system-measured information.

Many of our predictive dialer customers report that control of their outbound calling process is as important as the agent productivity gains. In these days of business process transformation, a tool that ensures adherence to the processes proves invaluable. Knowing that workload will be consistently processed in the prescribed priority permits management to concentrate on what's really important — improving the business process.

CHAPTER SUMMARY

Looking at every business with customers, products, and services, it's apparent that without a doubt improved customer contact translates into increased business. If you examine your customer/prospect contact needs, using the concepts and examples described in this chapter, you may find that predictive dialing will provide big payoffs, especially in business-to-consumer contact. Let predictive dialing take you anywhere; opportunities are everywhere.

Chapter 3

The State of
Predictive Dialing Today

This book's title is evidence of the authors' intention to limit its scope to the fundamentals of predictive dialing. What is not so evident, is the difficulty of separating predictive dialing from the context of the call center as a whole. Producing and delivering telephone contacts, the primary mission of the predictive dialer, is now just one facet of a much larger process. Aside from the increase in marketplace vendors, predictive dialing, itself, has not changed much during the last five to six years. Nevertheless, business applications based on this technology continue to grow as part of the evolving corporate call center environment. Predictive dialing's role in the call center coupled with the rapid advancements in information technology (IT) promise an exciting future for the telecommunications industry. Despite our urge to talk about where predictive dialing is going, in this chapter we will focus on how it has evolved until now. Just keep in mind, it is an emerging process.

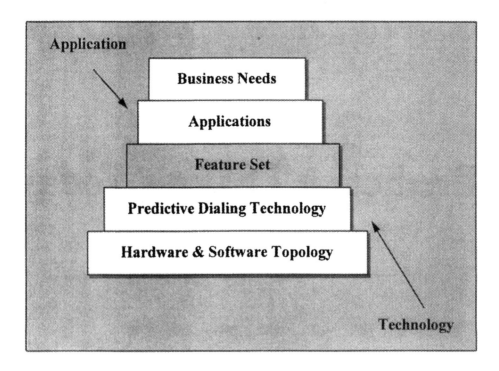

Figure 7
Technology & Application Perspectives

To characterize the state of predictive dialing, consider two perspectives: technology and application. The figure below illustrates a simple model from which to base these two perspectives.

Products develop from a mix of influences. On one end sales and marketing people identify business needs and petition applications. From the other end, technologists and engineers drive the underlying architecture. The result: the system feature set shown in the middle of this model.

Features are what users most look for and best understand. They are more easily compared than system architecture characteristics. However, it can be a perilous risk to go through the selection process and neglect the long-term strategic importance of underlying operating systems, messaging protocols and digital signal processing. Comparable features aside, all predictive dialers are not alike!

The development of predictive dialing systems and their markets has a similar history to that of a close relative — the PBX. Innovation seeded the early '70s competitiveness when the first non-Bell PBX vendors began entering the marketplace. Computer-controlled telephone switching was in its infancy and, thanks to software, a plethora of new features became possible. Early vendors competed primarily on features alone. But as more companies vied for a share of the business and the market matured, feature distinction began to blur. Today's PBX systems are packed full of features; most are never used.

Standalone predictive dialers are not as feature-rich as the PBX because of their more limited mission. But like the PBX, they have reached the point where major players advertise essentially the same capabilities. For this reason, only features worthy of distinction or unique to individual vendors will be covered in this chapter. In a "me too" feature comparison, what characteristics differentiate predictive dialers? To answer this question, we'll take a closer look, using the two perspectives in our model.

THE APPLICATION OF PREDICTIVE DIALING

PRODUCING AND DELIVERING THE CONTACT

The major promise of a predictive dialer has been to bring dramatic increases in agent productivity. So great is this increase, that when implemented to replace manual dialing operations, cost recovery is typically less than a year. Simply stated, the concept behind predictive dialing is to systematically deliver answered phone calls (contacts) and simultaneously present related information in a sequential, expedited manner to agents, without their need for intervention. This is accomplished in part by a technique called pacing.

Hang around any predictive dialing gym and someone is certain to utter the term "pacing algorithm." These two words describe the closely guarded recipe for controlling the speed at which contacts are produced and delivered. There are as many implementations as there are vendors. However, two basic ingredients go into all pacing algorithms:

Application

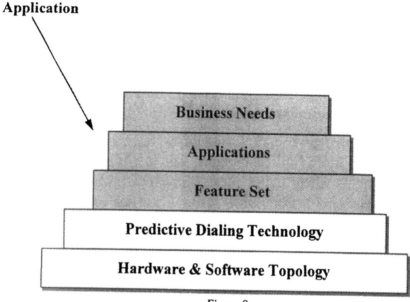

Figure 8
Application

- Projecting the time an agent will spend on an average call

- Dialing ahead on enough telephone lines to ensure at least one call attempt will result in someone answering at the exact moment an agent is ready for his/her next call.

Outbound dialing on more lines than a call center has agents is similar to the airlines overbooking passenger reservations. Statistically, there will always be a percentage of no-shows/no-answers. Of course, in predictive dialing, the trick is to make just enough calls to guarantee work for every agent as they wrap-up their previous call. When pacing is accelerated, an agent's talk time can increase to fifty minutes or more out of every hour. However, pushing the system to its limits to obtain maximum productivity is not without its price.

Using the airline analogy, there is always the possibility that more passengers will show up than there are seats -- something that can lead to some very unhappy customers. Here's where different philosophies surface relative to those two main components of pacing. The differences are apparent in both the vendor's approach to solving the problem and the user's quality-of-service objectives.

PREDICTING AGENT AVAILABILITY

The predictive dialer's precision in projecting the time agents will complete their calls depends entirely on the system's use of statistical averages stored in tables called histograms. The user's options for choosing a method for averaging agent talk times varies in predictive dialing systems. Some methods are unique to specific vendors.

For example, a common method uses a system-wide average talk time of all agents. This method works best when characteristics of all calls and the way they are handled are essentially the same. When dissimilar calling campaigns are concurrently active, a method that averages each campaign or group of agents separately is clearly better. Ultimately, since no two people or any two calls are exactly alike, a method that tracks each agent's own average statistics for each campaign yields the best result.

There are calling campaigns that have characteristics where the average talk time could be as short as ten seconds on one call, and as long as ten minutes on the very next call. How does this affect the averages that feed the pacing algorithm? Predictive dialer vendors employ their own techniques for defeating this circumstance. If the short call is the exception, one solution is to exclude short calls from the pacing calculations. Another solution is to permit direct stimulus from the agent. Who knows better than the agent when a call will end quickly or be extended beyond the average talk time? The stimulus is sent to the system from the keyboard or from a mouse click at the agent's workstation. This immediate feedback can reduce the wait time between calls or reduce the necessary number of calls launched by the predictive dialer to meet agent demand. The same sort of stimulus can be generated indirectly from an agent's workstation equipped with intelligence applications programs running on a PC.

GETTING A HIT

Life for a predictive dialer would be easy if every time it went to bat, it got a hit. Reality is, most call attempts made will result in busy signals, no-answers or answering machines. Even though these failed attempts to produce a contact are handled completely unattended by the predictive dialer, it must still make simultaneous multiple call attempts to increase the probability of at least one person answering, or one "hit." Hit rate is the term that describes percentage of calls answered out of total calls attempted. The higher the hit rate, the fewer simultaneous call attempts needed; the lower the hit rate, the more simultaneous call attempts needed to meet agent demand.

When dialing on multiple telephone lines for one agent results in two or more simultaneous hits, it's known as overdialing. All predictive dialers tuned for maximum productivity experience overdialing. It is how they counter low hit rates to increase agent talk time.

Predictive dialing systems vary considerably in how they dispose of or rerouting overdialed call attempts. The key is to maximize productivity while minimizing nuisance calls. A nuisance call is when a contact is established and no agent is available. The system must terminate or dispose of the call in some way. This is our airline passenger who shows up and no seats are left! There are options for eliminating nuisance calls that occur when the user's objective is to push the system for maximum talk time.

Typical alternatives to hanging up on the called party are playing a recorded message and putting the party on hold until an agent is available, or sending the call to an overflow or standby agent. It's a matter of balance, with potential nuisance being weighed against productivity gain. Telemarketing applications have fewer choices because restrictions imposed by many regulatory policies prohibit playing recorded messages.

One predictive dialing system offers a patented user-selectable feature that diminishes the possibility of overdialing. It does this by intelligently canceling other call attempts after a hit is detected when no other agents are available. This not only minimizes nuisance calls, but can reduce long-distance charges as well. Predictive dialer vendors and end users have differing viewpoints concerning customer sensitivity. The key is having the capability to seek a balance that makes sense for your business.

ELIMINATING NONPRODUCTIVE TASKS

The most non-productive part of making a large volume of calls is dealing with busy signals, no answers or bad numbers. Predictive dialing systems have varying degrees of success in accurately recognizing the various automatically processed signals and intercept tones returned by telephone companies. The techniques used today to monitor call progress are quite sophisticated and, in some cases, exhibit a high degree of accuracy in distinguishing between a live voice and an answering machine.

This is accomplished through pattern analysis of the cadence and frequency of the audio signal or voice energy. Some predictive dialing systems not only dispose of no-contact call attempts, but by monitoring the line after the number is dialed, they can also schedule appropriate numbers to be recalled at a later time. For example, call attempts resulting in a busy signal might be scheduled for recall within 2-5 minutes, while call attempts resulting in a no-answer could be rescheduled 30-60 minutes later. The system is responsible for tracking and reporting the result of each call attempt so the call center manager can update the calling record database.

UTILIZATION OF RESOURCES

In many predictive dialing applications, outbound contacts generate return calls. Since these inbound calls are related to the subject of the initial outbound call and generally require the same access to enterprise information, it is most cost-efficient to use the same agent work group to service these calls. Most vendors offer some facility for dealing with inbound calls. The capabilities vary from an inherent inbound call processing system to control of external PBX and Automatic Call Distributors (ACD) systems through computer-telephony interfaces provided by the major PBX/ACD vendors.

The choice, again, depends primarily on the business application. There are cases where the inbound call should take precedence over new outbound calls. In these cases, one may want to dynamically allocate agent resources within the same business unit to service the inbound load, filling idle inbound periods with outbound work. This can be achieved by a standalone predictive dialer utilizing the same central office facilities, without the need to interface with other PBX/ACD type systems.

Computer-telephony integration (CTI) is another method for cross-application utilization of agent resources and central office facilities. Currently, all major predictive dialing vendors offer CTI interfaces to some subset of the predominant PBX/ACD systems. Through external control (via CTI link) of the PBX/ACD switching system, the predictive dialing application software can change the role of agents from inbound ACD duty to outbound predictive dialing duty, swinging them either way to service both functions. The calls don't have to be related, presuming agents have been cross-trained.

Figure 9
Utilization of Resources

Distributing the work this way is an effective method of utilizing the agent's time, particularly because inbound and outbound peak traffic periods are usually inversely proportional. A challenge for predictive dialing vendors is producing consolidated reporting of inbound and outbound activities, since the extent of reporting data from the PBX/ACD is limited to what is available over the CTI interface or peripheral output ports. Another challenge is for such implementation to handle inbound/outbound calls on a call-by-call basis. The best advice on CTI configurations is to insist on industry standard interfaces when available, and stay away from proprietary methods, which may not survive the test of time.

MANAGING THE SYSTEM

Business needs and processes drive application design. Applications become the user's interface to control predictive dialer functions. System functions or features are designed to automate or assist in everyday user tasks. In a purely predictive dialing environment, administrative tasks fall into one of several general categories:

- System administration (dialing parameters and resource assignment)
- Information management (list management)
- Supervisory functions (agent help and monitoring)
- Reporting (both historical and real-time)

Predictive dialers have evolved from small five-seat configurations that used thousands of call records in flat text files, to systems with hundreds of agent seats on local area networks (LAN) using relational databases containing hundreds-of-thousands of records. As configurations grew, so did the need for good management tools. Today graphical user interfaces (GUI) have replaced character-based screens, and small stand-alone dialers are being replaced by larger distributed systems that can be centrally controlled. These are just some of the evolutionary changes induced by the ever-increasing use of automation to gain a competitive edge, significant advancements in digital technology, the pervasiveness of the personal computer and a business culture determined to re-engineer organizations.

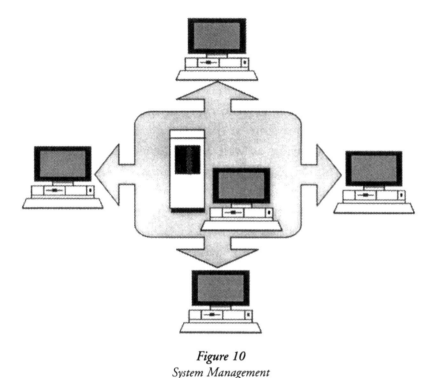

Figure 10
System Management

The design goal emphasis has shifted from merely increasing agent productivity to equipping the user with intuitive planning tools for developing execution strategies. Today, user-defined scripts hold:

- Definitions for triggers
- Alerts and alarms to provide real-time monitoring of agents
- Calling lists
- Dialing parameters
- And more

These events prompt actions by the system to self-adjust or correct operational parameters as conditions in the performance of the predictive dialer change or new strategies are invoked.

FOCUS AT THE DESKTOP

Over the last few years, we have seen desktop workstations replace the unintelligent terminal to access existing host-based applications. Earlier systems existed in a "dumb" terminal emulation environment, where all application logic would reside in the minicomputer or main-frame. It is clear that a $2,500 desktop workstation is capable of much more than replacing the character display provided by a $500 terminal.

In contemporary predictive dialers using client/server implementations, the low-cost processing power of the workstation has replaced some of the host processing and most of the application logic. Presentation of information as well as complex information management applications are handled entirely by the workstation, with much better performance and enterprise-wide information accessibility. This distribution of function and data is transparent to the user. Benefits gained by the advent of the desktop workstation and client/server technology are in the speed and ease at which enterprise-wide information is retrieved and presented to the agent.

Increasingly more powerful, less expensive desktop hardware is available. In 1982, Intel's 286 microprocessor was rated at 1 million instructions per second (mips) and cost $360. Compare that with Intel's 1993 launched Pentium. It cost $950, but pumped out 100 mips. So the per-mips cost was only $9.50 -- a 97% price plunge over 11 years. The microprocessor's memory capacity and speed are expected to triple in the next five years. Much of the processing load is being pushed out to the desktop.

Also, given their exposure to increasingly more complex software packages, users have become more sophisticated. A new generation of computer-literate users, with the skills necessary to take advantage of personal-productivity applications, populate the workforce Applications built on enterprise-computing architecture have shown added value and demonstrate a level of measurable benefit to the organization.

Products are available that let users manage their own application generation. The tools offer presentation, scripting, and enterprise-wide information accessibility features that don't require programming skills. Benefits include:

- Total flexibility in creating and editing applications
- Simplifying the interaction between agent and system through the aggregation of relevant information in a Single System Image View

- Navigation logic that analyzes agent input to control the conversation path.

Organizations continue to move toward employee empowerment and work group computing. The desktop has become the critical technology element in running the business. Workstations now provide power and capacity equal to that of mainframes only a few years old. Dumb terminals will continue to exist and be supported to some degree. However, the future lies with the desktop PC and its inherent client applications.

THE TECHNOLOGY OF PREDICTIVE DIALING

Functionality and feature content between predictive dialer vendors typically vary somewhat in:

- Ability to provide real-time and historical management information
- Flexibility allowed the user in controlling and monitoring system activities

Beyond user features, however, is the ability of the different predictive dialers to work seamlessly in the environment in which they are used. Here the difference in predictive dialing systems can be more dramatic.

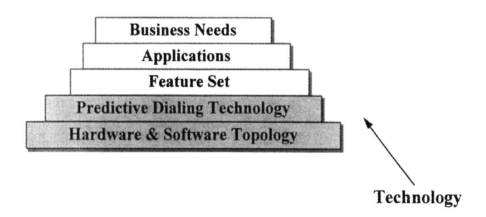

Figure 11
Technology of Predictive Dialing

The fundamental design principle of the underlying architecture should be one of open systems standards. Open systems architectures allow integration with legacy mainframe applications and access to diverse information sources over local and wide area networks. Standards will help protect existing and future investments. Next to people, information may be an organization's most valuable asset. Faster access to information improves customer service and differentiates a business from its competitors.

ENTERPRISE COMPUTING

Look for contemporary predictive dialers built on top of an enterprise-computing architecture. This is essential to meet the challenges of interfacing to other systems and provides a framework and migration strategy that will support new technology and continue to mirror organizational changes. Traditional predictive-dialing architectures have consisted of interfaces to mainframes, and dumb terminals in a single data processing center. This, in contrast to an enterprise-computing architecture, which is composed of multiple nodes of smaller computing workstations and servers that can be interconnected seamlessly across great distances by wide area networks (WAN).

There has been a slow migration by some predictive dialing vendors from traditional mainframe-centric computing to modern client/server architectures. These allow client applications developed by your organization (if desired), off-the-shelf applications and predictive dialing supplier's applications to work together within the same PC workstation. Where you can use tools familiar to you to run the system from any properly configured PC workstation. Don't be fooled by implementations of client/server where all the work (applications and processing load) reside exclusively on the server. This emulates the monolithic mainframe-centric configurations that client/server was designed to defeat. True client/server implementations distribute the processing load between clients and servers. Client-based applications and distributed computing environments will eliminate a lot of the headaches associated with application and systems integration. Put your future in the hands of suppliers that have adopted an open systems distributed client design.

A STATE OF TRANSFORMATION

Call centers are heading toward ever-increasing integration of inbound and outbound call processing; more sophisticated, customer service-oriented front end applications; and widely dispersed enterprise resources and information sources. Impetuses for this trend are a combination of technological advances, environmental factors and a rejuvenated competitive awareness among businesses for the need to enhance customer service.

The authors of this book have a clear vision of empowering the user with a distributed desktop workstation technology that will maximize use of multimedia to enhance the productivity and humanness of the call center environment. The infrastructure to accomplish this involves converging telephony and data services at the desktop to combine voice, text, image and video information. Information is only as useful as one's ability to gather and present it when needed.

CHAPTER SUMMARY

The role of predictive dialing within the call center and the rapid advancements in information technology promise an exciting future for this segment of the telecommunications industry. The state of predictive dialing rests on a threshold between increased productivity and technological changes that will transform cold call centers into warm Customer Care centers by the way information is stored, managed and presented. It is important for you to "build your house of bricks" with an infrastructure that will withstand the change of time.

Chapter 4

Case Studies

In this chapter we will describe predictive dialing implementations in five different industries: Banking/Financial, Cable, Healthcare, Utility and Services.

As you will see, each application deals with somewhat different issues: credit/debt collections, cable subscriptions, prescription-refill reminders, lawn care sales and protective services. However, all applications have a common need — contacting as quickly as possible, thousands of people and providing service over the phone.

We also analyze two ways a predictive dialing system could benefit a hypothetical (but typical) company. This explains:

- A simple method of calculating the call volume that existing agents could handle
- Similar calculations to determine the possible personnel reduction

Many growing call centers retain their current agent staff and dramatically increase their contact volumes which increases their business potential. Let's look at one example:

HYPOTHETICAL COMPANY

Table 1 outlines a company's current production based on 20 full-time telephone agents performing manual dialing applications 7 hours a day, 5 days a week. This amounts to 20 agents x 7 hours x 5 days or 700 agent hours.

Table 1

Type of Calls	Number of Attempts	% Distribution	Time with Customer & Wrap Up (per call)	Manual Dialing Overhead (per call)	Total Agent Time (per Call)
Right Party Contact	2,800	20%	3.0 minutes	2.0 minutes	5.0 minutes
Wrong Party Contact	1,400	10%	1.0 minutes	2.0 minutes	3.0 minutes
Answering Machine	1,120	8%	0.5 minutes	2.0 minutes	2.5 minutes
No Answer	7,000	50%	0.25 minutes	2.25 minutes	2.5 minutes
Busy	700	5%	0.25 minutes	1.75 minutes	2.0 minutes
Telephone Co. SIT Tones (3 Tones)	700	5%	0.25 minutes	1.75 minutes	2.0 minutes
Other Conditions (High & Dry, fax, modem)	280	2%	0.25 minutes	2.25 minutes	2.5 minutes
Totals	14,000	100%			

(Notes)

Right Party Contact	Speaking with desired person
Wrong Party Contact	Speaking with someone else at the called number
Answering Machine	Agent leaving a 15-second message and a 15-second wrap-up
Other	Fax modem calls, dead air (agent spends 15 seconds updating customer's call record)
Manual Dialing	
Overhead Includes	Time to retrieve a customer's record
	Time to review a customer's record
	Time to get a dial tone
	Time to dial a number
	Time to wait for call result (i.e. no answer, busy signal, right party answers, etc.)
	Agent rest time between calls

This data was gathered over a calling week. Your reporting systems may not provide some of this information, however, it is needed to assess what a predictive dialer will do in your environment.

1. The first information needed is the amount of time that all 20 agents spend making outbound calls. Lunch, breaks, meetings, absences and all other extra activities must be excluded. In our example, the agents are actually present for an 8.5-hour day. An hour lunch and two 15-minute breaks reduces this to 7 working hours. Now compile and total this information for all 20 agents during the measured period.

2. Tally the results by Type of Calls. In our example, 20 agents manually made 2,800 Right Party Contacts. This calculation

must be done for each of the seven Type of Calls categories listed. Now add up all these values to arrive at the total number of call attempts (14,000 in the example).

3. Next compute the Percent of Distribution column for each category. For example: 2,800 Right Party Contacts divided by 14,000 total call attempts equals 20%. This information will be needed later.

The next two columns in Table 1 will be the most difficult to determine.

4. The value in the Time with Customer & Wrap Up column for Right Party Contact is 3.0 minutes. The total number of minutes spent by all 20 agents with the Right Party was 8,400. The 3.0 minutes average per call was arrived at by dividing those 8,400 total minutes by the 2,800 calls.

5. Manual Dialing Overhead is determined the same way. For Right Party Contact the total for all agents for this function was 5,600 minutes. This divided by 2,800 equals 2.0 minutes average per call. The items included in Manual Dialing Overhead are listed under NOTES directly below Table 1.

You may ask how you can accurately gather and compile this data. It isn't easy unless you have an automated system. If you don't, then an estimate may work. But it won't be easy, because each agent performs in a different way for a different amount of time. Do your best.

6. Total Agent Time is the sum of the two previous columns (Time with Customer & Wrap-up and Manual Dialing Overhead). These totals can be used to validate your statistics. Multiply each of the seven call types by the Total Agent Time per call and then add these seven results. This number should equal the total agent time spent on outbound calling during the measurement time period. In our example, the time should total 700 hours (20 agents for 35 hours). Try it. You'll find it is pretty close.

USING A PREDICTIVE DIALER

Having listed the distribution of call results and their corresponding handling times in minutes, let's now calculate the productivity these 20 agents can achieve by using a predictive dialing solution.

Since most people prefer to take a few seconds between calls, let's assume each agent will spend a conservative 45 minutes per hour speaking with customers, leaving messages on answering machines and wrapping up transactions.

The percentage distribution of all calls in Table 1 was based on conditions that are not being processed and filtered by the predictive dialer. A new percentage distribution table must be created, after removing No Answer, Busy Signal, Telephone Company SIT Tones conditions and Other Conditions — all of which will be filtered by a predictive dialer.

Please note that Answering Machines can also be filtered out. However, in this example, we assume an agent will be leaving 15-second messages on answering machines and performing 15-second wrap-ups.

Right Party	.5263 times	3.0 minutes	=	1.58	minutes
Wrong Party	.2632 times	1.0 minutes	=	.26	minutes
Answer Machine	.2105 times	0.25 minutes	=	.05	minutes
Overall Weighted Transaction Time				1.89	minutes

Table 2

Table 2 is similar to Table 1 with some exceptions:

1. No Answer, Busy Signal, Telephone Company SIT Tones and Other Conditions have been eliminated since the telephone agent will not be involved with these call results.
2. Percentages have been recomputed based on the only three types

of calls the agents will handle.

3. Manual Dialing Overhead is zero since there is no overhead — the predictive dialer performs all the dialing.

4. The last column is a Weighted Average call-length calculation.

So what is the length of an average call when agents are using a predictive dialer?

To determine this, we must recompute the distribution percentage of each call type. The number of attempts of each call type during the measurement period depends on who or how the number was dialed. Had a predictive dialer placed the calls, the number of attempts would still be the same. The key difference in the above example is that the predictive dialer only involves the agent in the three types of calls. Add calls transferred to agents and their new distribution percentages become:

- 52.63% Right Party
- 26.32% Wrong Party
- 21.05% Answering Machine

A shortcut to calculate Overall Weighted Average Transaction Time is to multiply the percentage of the distribution of each call type by the Time with Customer & Wrap-Up time for that category and then total these results. Using Table 2:

The Weighted Average Transaction Time is the amount of time an average agent will spend on the average call. We can now proceed to the next step.

DIALER PRODUCTIVITY RESULTS

From here on, it gets easier!

We will determine the productivity increase when using a predictive dialer based on each agent being productive 45 out of each 60 minutes. The other 15 minutes allows rest time between calls.

First, by dividing our 45 minutes productive time by the Weighted Average Transaction Time of 1.89 minutes (Table 2), we get 23.81 calls that an agent will, on the average, process for each 60-minute work period.

Our example is for 20 agents working 35 hours or 700 agent hours per week.

700 agent hours
x 23.81 calls per hour
_____
16,667 calls processed.

Table 2 indicates that 52.63% of the calls processed are Right Party Contacts and 52.63% of 16,667 total calls equals 8,771. So in the same number of hours, the same 20 agents could produce 8,771 Right Party Contacts using the predictive dialer versus only 2,800 manually. Using the predictive dialer produced over three times the number of Right Party Contacts compared with the original manual dialing in Table 1.

In other words, this exercise demonstrates a 200 plus percent productivity improvement. Another important consideration: if 2,800 Right Party contacts is all you need (you don't wish to contact the same customers three times as often) then you could reduce your staff to 7 agents and achieve the same results as with the 20 manual dialing agents.

One last point. If we were projecting results using our predictive dialer, we would use over 50 minutes of agent productive time per hour for such calculations, while still maintaining a nuisance rate below 1%. This amounts to an additional 35% productivity increase while still giving the agents a 15 to 20 second rest between calls. Further, if you desire to combine inbound and outbound calls for some groups of agents (Dynamic Inbound/Outbound), additional productivity increases can be achieved.

Compelling isn't it!

CASE STUDIES

Uncovering information that relates to your industry provides an opportunity to expose issues and consider solutions. The following case studies profile companies by: industry, size, business applications, configuration, situation/needs and solution/benefits.

These case studies outline issues and offer solutions that call center managers encounter as they plan and manage their daily activities. Following are examples of how some other companies manage telemarketing, customer service and collections calling campaigns, and how they benefit from using predictive dialing solutions.

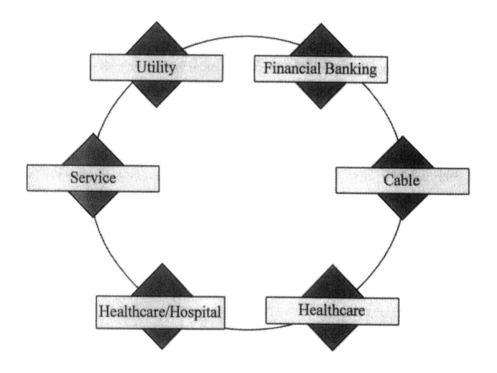

Figure 12
Case Studies Industries

CASE STUDY: BANKING/FINANCIAL

- **Size:** Millions of customers nationwide
- **Business Applications:** Credit card collection, consumer loans and telemarketing
- **Configuration:** 96 workstation positions

SITUATION/NEED:

One of the country's top ten banks performed a thorough review of their entire operation.

They decided, in part, to reduce direct/indirect labor and operational costs, and expand customer servicing via telemarketing. With the increase in their portfolio, they wanted to:

- Maintain their current staff of 21 agents in the collection area
- Centralize multiple collections units
- Create an environment where all agents could universally access information from multiple applications resources and simultaneously manage both inbound and outbound calls

The company already had a predictive dialer in place, but felt their goals could be better accomplished with a more comprehensive, enterprise-wide predictive dialer solution.

SOLUTION/BENEFITS:

Currently, 21 collections agents contact an average of 50,000 customers each month. This contact rate increased by 17% and promises-to-pay increased by 25% over the previous dialer's results. Considering the company upgraded from another vendor's predictive dialing system this represents a significant improvement.

Once the bank experienced this productivity gain, they quickly expanded into telemarketing using the same predictive dialing system. An additional group of 40 telemarketers contact over 120,000 customers a month across 25 states to promote a new cobranded gasoline company credit card. More recently, an additional 35 collectors have begun to manage the gasoline company accounts. Now a single predictive dialing solution supports a banking division with dynamic inbound/outbound automation, collections and telemarketing call activities.

With this system, call handlers and management experience benefits that range from increased employee and customer satisfaction to increased productivity. Management can track the productivity of each agent and receive reports by agent, campaign and inbound abandonment rates. In real time, they can track statistics and manage results on an ongoing basis. They're much better informed than ever before.

Another benefit of the predictive dialer system is the increased efficiency from the Windows workstation environment which contributes to reduced key strokes and increased call handler efficiency. Functions and steps are predefined. Switching to another screen or application is automatic. At a touch of an icon, call handlers can connect to a specific financial application screen, request a letter be sent and more. Every time a nonproductive task is automated, call handlers have the opportunity to increase their number of contacts and improve on customer response times.

Another system feature that has helped the bank to maximize agent productivity and increase customer satisfaction, is the system's ability to dynamically manage inbound and outbound calls. During slow inbound periods, outbound call volumes are automatically increased so call handlers' productivity is maintained. As inbound calls increase, outbound dialing is decreased to give inbound calls priority. The bank is able to optimize its staff, and customers calling in receive immediate attention.

When call handlers connect to an inbound caller, the system automatically updates the calling list. This customer won't be called unnecessarily via an outbound call campaign. This reduces nuisance calls. Before they installed a predictive call processing system, management didn't have the option to automatically update the customer database. Now they do.

One of the system's greatest benefits to the bank is the elimination of employee overtime. Call handlers choose their schedules, work in prime time, achieve their goals and maximize their incentives.

CASE STUDY: CABLE (CATV)

- **Size:** Thousands of subscribers across the South
- **Business Applications:** Telemarketing and customer service
- **Configuration:** 16 workstation positions

SITUATION/NEED:

Proactive customer contact programs are key in:

- Retaining a solid customer base
- Providing ongoing customer service
- Improving the efficiency and productivity of the operation
- Generating incremental revenue
- Controlling overhead costs

With ever-tightening federal regulations, it's even more important for Cable TV companies to provide improved customer service. That's why more and more cable companies are streamlining operations by installing predictive dialing solutions.

Before this cable company installed a predictive call processing system, they would split and process their calling lists manually. Sixteen telemarketing agents tracked their calls on "tick" sheets.

Pay-per-view, upgrade and acquisition campaign sheets were passed between telemarketers. Agents couldn't easily switch campaigns. Each agent could only make fourteen presentations an hour. Management had difficulty analyzing outbound activity. Agents spent more time getting prepared than actually talking with potential customers. The cable company realized they needed a solution to their call center problems.

SOLUTIONS/BENEFITS:

Management installed a predictive dialing system. They now have a clear picture of their daily call center activities and results.

With the predictive dialer, daily calling lists are automatically downloaded from their cable application host computer. As agents log on (typically in the evening hours), the system immediately starts dialing. Only live calls are connected to each agent. Based on user-established rules for each campaign, the predictive dialing system automatically simultaneously calculates the correct number of calls and the appropriate pacing. Agents easily manage different types of campaigns, switching from one to another with hardly a pause.

Frequently, the cable company's subscribers call in requesting immediate information about HBO and Cinemax channel promotions. It's important to treat these calls as a priority and respond to them as soon as they are received. And that happens, because their predictive dialing system dynamically manages inbound and outbound calls. As inbound calls arrive, the system interrupts outbound dialing. Inbound calls are given higher priority. When inbound calls decrease, outbound calls are increased. This seamless transfer of calls automatically takes place within the predictive dialing system and is transparent to the agents.

The company's telemarketing department also targets other campaigns: customer surveys identifying specific needs and acquisition of new customers. These calls are more focused, providing better customer communications. With real-time access to billing and historical customer information, they'll be able to immediately log customer orders, handle trouble tickets and much more. Additionally, one week of each month, agents make 3,000 reminder calls urging subscribers to make payments and avoid late fees and/or service interruption.

The CATV company increased sales 40% and paid for the system in less than a year.

To accommodate diversified subscriber needs, this cable company plans to upgrade their predictive dialing system software. The Universal Agent software will allow their agents quick and easy on-line access to other cable software applications from any properly configured PC workstation.

Different campaigns...different reasons to call...using the predictive dialer, the cable company constantly ensures quality contacts and a high level of customer satisfaction. Thanks to the benefits of the predictive dialing enterprise-wide solution, subscribers receive the service they want and the cable company controls its operational expenses.

CASE STUDY: HEALTHCARE

- **Size:** 1-million customers nationwide
- **Business Applications:** Customer service
- **Configuration:** 25 workstation positions

SITUATION/NEED:

Recognizing the growing significance of managed care, the nation's largest pharmacy chain created a special division to administer prescription benefits to over one million customers across the United States.

One of the leading causes of suboptimal health and increased healthcare costs is prescription noncompliance. According to the Task Force for Compliance, there are annually 125,000 unnecessary deaths and an $8.5-billion excess in hospitalization costs related to poor compliance. The new division's goal is to provide healthcare management services that reduce customer healthcare costs, improve overall health of members and ultimately maintain and increase the company's customer base.

From their beginning, this pharmacy chain has focused on customer service. That's why they chose a predictive dialing system as part of their solution. With the predictive call processor, the pharmacy division makes a significant difference in how patients follow through with prescription therapy. Customer service representatives call patients and inquire about their prescriptions. They also encourage them to take critical medications, and to renew their prescriptions on time.

SOLUTION/BENEFITS:

Currently, 25 customer service representatives make 40,000 presentations every month. When capacity is reached, they anticipate expanding to 96 representatives, more than tripling their calling capacity to 140,000 contacts each month. After the first year of operation, the division expects to have 174 stations and reach well over two million patients a year. The predictive dialing system can expand to over 256 stations.

The predictive call processing system uses the data stored in the company's central database and runs applications that ultimately touch the lives of hundreds of thousands of patients. Customer service representatives identify patients on chronic medications, call to inquire about their prescription therapy, encourage them to take critical medications on time and remind them to renew their prescriptions as directed by their physicians. Patients know the pharmaceutical company is always there to assist.

Representatives benefit because they control the length of each call. If they need more time with a patient, they touch the Extend Key which signals the system to reduce outbound calling. By extending calls, agents can spend the necessary time with each patient. Patients receive the appropriate follow-up information and agents don't feel rushed.

Of the 64,000 patients reached by the 25 customer service representatives during an initial three-month period, almost 41,000 agreed to have their prescriptions refilled. A sampling of patients receiving reminder calls is contacted monthly to determine the percentage of patents who actually pick up their medications. Results show that 92% picked up their prescriptions at the pharmacy. According to industry standards, this surpassed the anticipated results by 20%.

As more patents follow their prescription therapy, expect a reduction in excess hospitalization costs.

CASE STUDY: HEALTHCARE/HOSPITAL

- **Size:** 2.3 million clinic visits annually
- **Business Applications:** Patient service and collections
- **Configuration:** 10 workstation positions

SITUATION/NEED:

As a major medical center, this hospital has in excess of 36,000 annual patient admissions. Additionally, the hospital's health system's main revenue source comes from outpatient services provided at 40 specialty centers located on the hospital campus, as well as 37 full-service outpatient satellites strategically located throughout the metropolitan area. Due to its inherent size and revenue producing capabilities, significant emphasis is placed on the collection of personal balance accounts.

Because outpatient clinic visits total over two million annually, the continued development of cash collection strategies to minimize bad debt write-offs is essential.

The hospital set up specific actions to handle these needs:

- Agent training
- Collection letter programs
- In-house collections programs
- Small balance reviews.

Programs developed to achieve these goals place a great deal of emphasis on patient contact and communication regarding personal balance amounts and past due liabilities. The hospital still wanted to better manage the extraordinary volume of small balances which result from processing 2.3 million clinic visits a year. They chose to move to new, more efficient technology.

SOLUTION/BENEFITS:

The decision to purchase a predictive dialing system was based on account volumes, cost analysis and projected cash returns. The hospital's call center now has ten agents who target a large number of small balance accounts that:

- Aren't covered under the review criteria of the large-balance in-house collection staff
- Are overdue and are to be systematically referred to an external collection agency

The hospital, which has 26 monthly payment postings, downloads this data by overdue age and dollar amount. Part-time agents make calls during the day, and an evening crew works from 4:30 to 8:30 p.m. In every instance, the collections group works in tandem with customer service.

It's easy to make a high volume of calls, but when patients raise questions about insurance or other issues, a process needs to be in place to resolve these issues. Now they have this process.

Agents call the patients with small balance accounts only after those patients have received three statements. When there's an outstanding insurance issue, the agent uses the Help Agent feature on the dialing system to forward a call to a Customer Service Representative. Processing these calls, ten agents collect approximately $75,000 a month.

About 3,000 accounts are downloaded twice a week with each agent communicating with about 35 patients an hour. Agents separate the calls under four categories:

1. Promise to pay
2. Call customer service
3. Refused to pay/refer to outside agency
4. Bad number/requires skip tracing

Because providing additional customer service is important to the hospital, courtesy calls are made to self-pay accounts where insurance registration and address data are checked. They also call patients to remind them about appointments — verifying, canceling or rescheduling. The ultimate goal is to preregister patients, verifying insurance/address and indicating any copay and deductible amounts.

Additionally, prior to the use of the predictive dialing solution, 40 to 50% of customer calls on small-balance accounts were related to front-end registration/address errors. Now with call processing automation and PC workstations and software, such errors are reduced to only about 10%.

CASE STUDY: PROTECTIVE SERVICES AND LAWN CARE

- **Size:** 1.5-million customers nationwide
- **Business Applications:** Telemarketing, collections and customer service
- **Configuration:** 40 workstation positions

SITUATION/NEED:

This major corporation operates two divisions that needed help with their telemarketing, collections and customer service. They chose a predictive dialing system to help them increase productivity and provide better customer service.

1. Protective Services Division:

(Security Systems and Pest Control)

Management wanted to increase productivity while at the same time proactively contact customers to provide solutions to some of their most pressing maintenance and security issues. Their goal: provide best-in-class service.

They chose the predictive dialer because it:

- Streamlines their operations
- Delivers real-time reporting
- Provides inbound/outbound call handling
- Offers call back capabilities
- Contributes agent and group pacing
- Supplies automatic zip-code adjustments

SOLUTION/BENEFITS:

Customer service representatives process thousands of phone contacts each month and ask, "Are you satisfied?" type questions (regarding newly installed security systems), before customers receive their first bill.

If the customer has a problem, account information or concerns are automatically sent to one of 48 branch offices and the customer's issue is given top priority. It's to the advantage of both customer and company to be proactive. Customers receive the service they deserve. Ultimately, the company experiences reduced cancellation rates.

Each month collection agents contact approximately 20,000 customers to discuss past-due accounts and payment options. The Pest control Department experienced an 87% increase in revenue and the Protective Services Department 57%, yet they only added 25% additional staff. The company gained a 141% increase in renewal revenues. This demonstrates the success of the predictive dialing system.

2. Lawn Care Division:

SITUATION/NEED:

The Lawn Care Division's mission matches the Protective Services division. The primary goal is providing quality service.

Teleservicing representatives (TSRs) spend the time necessary to ensure each customer is completely satisfied. These calls often take longer yet offer an opportunity for representatives to provide options to prospects and customers.

The Lawn Care Division runs three or four different campaigns a week based on the agronomic seasons across the country. Flexibility to call one region in the morning and switch to another in the afternoon is critical.

SOLUTION/BENEFITS:

Sixteen TSRs contact about 15,000 installed-base customers weekly. They provide additional services or upgrades and generate 800 to 1,280 sales contracts. Four TSRs initiate 5,000 cold calls (25 - 30 contacts per hour) and produce approximately 250 new sales weekly. Real-time reports provide current statistics so adjustments can be made in calling patterns. This means TSRs can reach and make more presentations to customers each hour.

The predictive dialer provides flexibility to meet the needs of each division. Collection agents contact protective-service and lawn-care customers. TSRs cross-sell products/services for many divisions from pest control to plantscaping.

As customer survey campaigns touch off inbound calls, call handlers are trained to process these and respond to customer requests. This call center is set up to dynamically manage inbound and outbound calls. As inbound call volume increases, outbound calls are decreased. In reverse, as inbound call volume drops, outbound calls are increased. Inbound customer calls are given priority over outbound calls. Agent productivity is maintained, and all of this happens automatically without a supervisor's intervention.

It's a unified effort for each division to conduct business in its own individual way. Yet, the primary goal is universal: ensure the highest level of customer service. The predictive dialing system helps meet that goal.

 ## CASE STUDY: UTILITY

- **Size:** 600,000 customers covering 31 counties
- **Business Applications:** Collections
- **Configuration:** 6 workstation positions

SITUATION/NEED:

Utility companies face numerous challenges. Foremost is meeting the growing utility demands of the American people by providing reliable service at a reasonable cost. Others include:

- Increasing revenues by decreasing uncollectibles
- Reducing costs associated with disconnects
- Improving customer service.

How do utilities go about collecting on past-due accounts in a cost-effective manner? Millions of dollars are lost each year due to late-paying customers.

One electric company needed to reduce overhead yet gain an effective method of collecting on past-due accounts. In their geographic area, one of every six customers was past due paying their electric bill.

The utility's collection agents manually dialed customers and due to the time this took, were unable to complete their daily calling lists. So, often no personal contact was made. The utility would only send letters. They needed a way to speed calling and boost collections.

SOLUTION/BENEFITS:

The utility chose a predictive dialing system for residential collection applications. The collection department downloads 6,000 names nightly to the predictive dialer and, within ten hours, calls all 2-month-past-due residential accounts with outstanding balances of $100 or more.

They needed a flexible system so users could establish dialing parameters and create customized calling lists. With the predictive dialer in place, the collection department categorizes calling lists and strategies every day. Not only did productivity increase, but they also gained access to other software applications.

The company chose a predictive dialer solution because they needed to be more competitive and lower their costs. The predictive dialing system provides on-line, real-time updating. Collection agents can hot-key between applications and gain access to the most accurate, up-to-date past-due account information. As collectors talk with customers, they connect directly to the host applications desired and process new account information.

The benefits: increased revenue and improved customer service. Agents have experienced more than a 100% increase in customer presentations and a 60% increase in promises-to-pay. Plus, by making payments when they're only one or two months late, customers avoid service interruption.

To ensure positive customer contacts and adhere to state and federal regulations, agents also use Preview Dialing. The law requires that customers who have overdue accounts be called for nonpayment 72 hours prior to service termination.

Preview Dialing lets agents review the customer's current billing status (i.e. has a recent payment or arrangement to pay been made?). By previewing a customer's record, agents avoid making unnecessary calls that could prove embarrassing.

Unattended dialing is another predictive dialer function that this company regularly uses. As a service to some 40,000 customers who are eligible for low-income, home energy assistance, the utility downloads this specific customer list and the predictive dialing system automatically dials and leaves a recorded message notifying them of their energy assistance. The predictive dialer completes such calls and message delivery totally unattended (no agent involvement). The utility has received considerable funding because of this informational program.

Dynamic inbound/outbound dialing works in conjunction with preview and unattended dialing to help the company capitalize on each customer contact. The system captures inbound calls and automatically immediately routes them to an available agent. Because of the dialer, the utility talks to the largest number of customers possible and gains more promise-to-pay commitments.

Automated telephone call processing, which is not new to the utility industry, has grown in popularity now that the technology is more affordable. For this utility, the predictive dialing solution paid for itself in less than one year.

CHAPTER SUMMARY

This chapter truly shows the power a predictive dialing solution provides. Since every business needs to reach the Right Party as often as possible (reminders, relationship building, telesales, customer care calls, etc.), retaining current staff instead of reducing it could pay big dividends over time.

FINAL NOTE

A Call Statistics Worksheet is included in the Appendix to help you collect call processing information needed for various calculations. Additionally, blank copies of Table 1 and Table 2 are included so you can calculate your own results.

Chapter 5

What to Look for When Choosing a Predictive Dialer

The proceeding chapters described what predictive dialers are, how they are used, where predictive dialing systems are today and how through simple calculations, to determine if you need a predictive dialer. This chapter explores what you need to know before purchasing a predictive dialing system.

Are all predictive dialers alike? Not really. Predictive dialers are similar in the same way cars are similar. All cars provide the same basic benefit, transporting people from one place to another. How fast and how comfortable the ride — depends on the car.

	Attributes Common to Four-Wheel Passenger Vehicles	Differences Between Four-Wheel Passenger Vehicles
Have engines	(motive power)	Different size engines
Can be steered	(operator control)	Power or manual steering
Have seats	(capacity)	Different number of seats
Have doors	(ways in and out)	Different number of doors
Go forward and backward	(versatility)	Front- rear- or four-wheel drive
Have warning indicators	(alarms, alerts)	Different kinds of indicators
Can be turned	(change of direction)	Different turning radius
Have licenses	(approvals)	Different types of licenses
Have locks	(security)	Different kinds of locks
Have brakes	(way of stopping)	Different kinds of brakes
Get us from point A to point B	(travel on highways)	Different ease of operation
Have mileage ratings	(fuel efficiency)	Different mileage ratings

Using the car example, let's more closely examine the similarities and differences between predictive dialers.

It's interesting to note that seven of the twelve similarities are also listed as differences. Luxury, aesthetics and, of course, price are also important factors. Because the cost of using a predictive dialer is such a major consideration, it's the first criteria we'll examine.

WHAT ABOUT THE PURCHASE PRICE

Let's again use the car example. What's more important: the initial cost of the car or the total cost of owning and operating the car over its useful life? Compare two similar cars, one priced at $12,000, the other at $20,000, and examine what the total cost of owning each is over a three-year period.

The $12,000 vehicle gets 10 miles per gallon while the $20,000 vehicle gets 30 miles per gallon. Now if we drive 100,000 miles over three years at a cost of $1.25 per gallon, the total costs are:

	Vehicle A	Vehicle B
Original Cost	$12,000	$20, 000
Cost of Fuel	$12,500	$ 4,167
Total Cost over 3 Years	$24,500	$24,167

These projected costs of owning each car are close, but somewhat surprising in that the more expensive car actually costs less to operate over a three-year period.

In evaluating predictive dialers, even relatively small differences in their capabilities can result in tremendous differences in overall value to the customer.

When comparing the cost of predictive dialers to the cost of owning a car, the fuel cost is the cost of telephone agents. Most industry survey data indicates that the fully-loaded cost of a telephone agent varies between $20 and $35 per hour. To be conservative, we'll use the lower figure of $20 per hour. This makes the annual cost of a full-time $20/hour agent, $41,600 per year.

Now if one predictive dialer produces 5% more productivity per agent, then the same amount of work could be done with fewer agents (or more work could be done with the same number of agents). This means 38 agents could produce the same results as 40 agents. So during a three-year period, the more efficient predictive dialer saves two agents per year, or six agents. This amounts to a personnel savings of approximately $249,600.

The difference in purchase price between the two predictive dialers would have to be at least that much for the less efficient system to be an equal. This is over $6,000 per agent position. Remember, we assumed only a 5% difference in productivity. Since most predictive dialers have a useful life of four or more years, at the end of four years, the difference in price per seat to break even would be nearly $8,500! Put another way, you should be able to pay at least $6,000 more per seat for the more expensive predictive dialer and still save money.

So, the first what-to-look-for when selecting a predictive dialer is not cost, but life-cycle cost. Buyers can't be distracted by the initial cost.

NUISANCE CALLS

While we're talking about productivity, we need to look at another important factor: nuisance calls. As outlined in preceding chapters, a nuisance call occurs when the predictive dialer processes a call that is answered by a customer, but no agent is available to take the call. The predictive dialer can then either disconnect the call, or place the customer on hold.

Now, the real struggle for call center managers is to find a happy balance between keeping agents at maximum productivity and keeping nuisance call generation at a minimum.

So, what can the manager do? In Chapters 2 and 3, we talked about companies that place the caller on hold, using recorded messages until the agent comes on line. Unfortunately, with this procedure the caller will most likely be hostile when the agent finally takes the call.

The only other way to handle nuisance calls is to immediately hang up as the person answers the phone. This creates another problem. Many people, especially people who live alone or are home alone, become terribly frightened when they answer the phone and the caller hangs up on them. This technique is not a good solution and should be avoided.

Is there an acceptable balance between increased telephone agent productivity and nuisance calls? Each company must decide where their standards are in relationship to this issue. Beware of predictive dialer vendors who claim their products will never create a nuisance call. Gumperson's Law applies here: "What can happen will."

Many techniques are recommended to minimize nuisance calls. One frequently offered is referred to as Agent-on-Demand or Overflow Agents. This option, which has some merit, has a number of telephone agents (frequently supervisors) available (waiting) to accept these over-flow calls. There are two scenarios offered for this method:

1. Some dedicated agent(s) sit(s) and wait for the overflow condition to occur. This is like having a spare agent(s) who is (are) really unproductive but it does solve the problem. But is a company willing to pay $40,000 per year for each of these agents?

2. Supervisors are the spare agents. However, if they are acting as Agent-on- Demand type agents, they really can't do much super-vising because they must also be in the agent mode to handle the overdialed calls.

When choosing the right predictive dialer, the buyer should have a clear understanding of how it handles nuisance calls. How will these calls be minimized? How will they be handled when they do occur? Since we know they can't be eliminated (Chapter 2 talks about over-dialing), the key is knowing the capabilities of the predictive dialer to minimize and manage nuisance calls.

In one customer application, we have been able to fine tune their system so that it produces seven nuisance calls out of 30,000 complet-ed calls. Even this extremely low percentage still produced 47 minutes per hour of productive talk time per agent.

A general rule: less than a 2% nuisance call rate should be achievable without overflow agents. The exact results, however, are dependent upon hit rates, talk times and other factors.

BASIC MINIMUM REQUIREMENTS

Predictive dialers all perform the same basic function: automatically dialing and locating desired parties. Predictive dialers differ, however, when it comes to how they perform that function, how efficient they are in performing that function and how easy it is to manage the system. Every predictive dialer vendor will have slightly different criteria for evaluating their solution. Usually, the criteria they use definitively shows that their predictive dialer is the best product.

What the buyer finds, however, is that predictive dialers have different strengths and that no predictive dialer does everything better than the others. So let's look at some of the basic requirements and capabilities that all predictive dialers should have.

Another important factor when selecting a predictive dialer is being able to establish and control its behavior. In other words, a predictive dialer that has lots of features is great, but if it doesn't work the way the buyer wants, it probably isn't the right choice. The predictive dialer must execute telephone contact strategy based on the buyer's specific business needs and requirements.

A FEW DEFINITIONS:

To better understand the criteria required in selecting the right predictive dialing system, we need to define specific terminology:

HIT RATE

Hit rate means the odds that a phone number called will be answered. Percentages of hits vary greatly by time of day, day of week and quality of calling lists.

APPLICATION

A thirty-day delinquent credit card calling is an application. A thirty-day delinquent personal loan calling is a different application. A "Thank you for your first order." calling campaign is yet another application. An application can consist of multiple calling lists (typically with same/similar characteristics).

CALLING LIST(S)

These are lists of customers and their phone numbers that the predictive dialer will call. Each calling list is related to an application, and each application may have multiple calling lists. A 30- day delinquent credit card application might have three lists: over $500 past due, $200 through $499 past due, and less than $200 past due.

PACING

A usually proprietary set of the computer logic that makes decisions on when to make additional phone calls. Pacing uses the hit rate and other internal statistics (application, agent talking, time of day, etc.) to determine how many calls to make. Each supplier has its own style of pacing.

PREVIEW DIALING

An operating mode where the agent, before the number is dialed, is presented with the screen data about the party to be called. The dialer places the call and then connects the agent to the ringing line. This is on automated (not predictive) dialing.

THE BASIC MUST HAVES

When selecting a predictive dialer, there are certain basics the system must have. These features include:

PACING LOGIC

The manager must be able to set a different pacing mode for each of the running campaigns.

HIT RATE TRACKING LOGIC

The predictive dialer must allow automated activities: load new calling lists, disable no-answer recall logic, enable call backs, etc. based on hit-rate thresholds set by the user.

CALL STRATEGY MANAGEMENT

We already defined the terms Application and Calling List(s). Applications are sometimes called campaigns, buckets, gates or splits. In essence, the term application is an aggregate of these.

Modern outbound contact schemes frequently use some type of

scoring system to arrange all names into a sequence from most important to least important.

In collections, this might be a sequence based simply on the amount of money past due, from most to least. In customer service, it might be based on a complex system that considers date of last contact, volume of business, number of years as a customer, time of day, etc. In telemarketing, the strategy might range from best-fits-the-profile to least-fits-the-profile of contacts interested in a product or service.

We may want to further separate this sequenced contact into more specific groups. For example, in collections, the grouping could be: 1) balances $500 or over, 2) more than $200 but less than $500, and 3) less than $200 past-due balances. These groups then become calling lists.

Why do this? Frequently the outbound calling load exceeds the ability of available agents to process calls. By dividing the work into calling lists, a manager can make sure that the most important work is done first.

The predictive dialer must support such multilist, multicategory automated list preparation in process. This process must be self-started based on date availability and other various triggers.

FULL UTILIZATION

Using our collections example, here's another way of looking at this: As long as all agents assigned to this application stay fully utilized contacting people with balances above $500, should they even try calling those with lower balances?

The key statement is fully utilized. When agent productivity starts to fall, it's time to start activities on the next most important list. So it is imperative that the predictive dialer reacts to the user-established rules of when to add the next critical calling list.

TRIGGERS-BASED RULES

The predictive dialer must offer methods that automatically (without human intervention) change the work load assignments as necessary. User-defined triggers do this. Triggers could be: time of day, percentage of penetration of a calling list, or a hit rate decreasing below a certain level. Triggers should:

- Start an additional list
- Stop a specific list
- Start certain types of calls within a list (start attempts on all call backs for the day)
- Stop certain types of calls within a list (stop all busy, no-answer recalls)
- Allow multiple lists to be active simultaneously
- Allow definition of the thresholds to change
- Allow multiple applications to run simultaneously, each with its own triggers
- Cause notifications, alarms and other activities to be started

With this capability, the manager prescribes the rules and the predictive dialer follows them regardless of day-to-day or time-to-time variations in workload or number of telephone agents assigned.

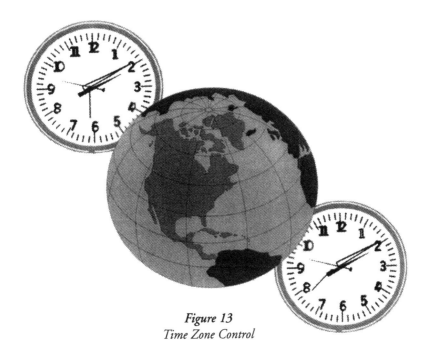

Figure 13
Time Zone Control

TIME-ZONE CONTROL

The U.S. has four primary time zones. Managers should be able to automatically define local times of day for starting and stopping calls in each time zone.

CALL-BACK DATE AND TIME

Frequently on a completed call, the person desired is unavailable. It is important that the predictive dialer allow the agent to enter a suggested date and time to call again. If the time is for the same day, the dialer must make sure this call occurs at the appointed time. If the scheduled call is for a subsequent day, the dialer, in conjunction with MIS, must give this same assurance.

RETRY FOR NON-CONTACTS

Many call attempts result in fax/modem, busy, no answer, various types of three-tone signals or answering machines. The manager must be able to use the predictive dialer to define the retry rationale for each result. This rationale usually includes how often and how many times a number will be redialed. A busy signal indicates someone is there. So retry in 5 to 15 minutes makes sense. Conversely, a no answer means no one is home; therefore, a retry in 2 or 3 hours might make sense. The predictive dialer should also allow the manager to change these rules during the calling sessions. For example, at 6 p.m., it might make sense to change the no-answer retry time to 30 minutes. As defined under Triggers, all such conditions should be automated as part of overall call-strategy rules. Answering machines are a special case and will be covered next.

ANSWERING MACHINES

Depending on the application, a manager may or may not want answering machine calls connected to agents. Sometimes, when a customer-specific message needs to be left, answering machines can be treated the same as live human answers.

The manager may want to screen these calls (not connect them to agents) and leave an automated recorded message, or screen these calls and not leave any message, plus delay retrying calls until a later specific time.

This approach allows agents to speak directly to the customers, once they arrive home. This more personal approach reinforces Customer Care.

One more note about answering machine detection. It is not an exact science! Remember the predictive dialer is trying to differentiate between a live answer and a recording. It will occasionally make errors both ways. No 100% accurate method exists. If the logic is tuned to be very accurate on detecting answering machines, the number of errors on classifying live answers as machines increases. Avoid errors in this direction as they appear to the parties called as nuisance calls! Tune the logic to be very accurate in detecting live answer and accept some errors in classifying machines as humans. This way agents can quickly terminate error calls with minimal time loss. The live answer classification has a near 99% accuracy and the machine classification an 80 to 90% accuracy.

Figure 14
Call Back Queue

PREVIEW DIALING (SEE EARLIER DEFINITION)

This mode is quite useful in training new agents. The predictive dialer should permit the manager to specify any agent or group to operate in the preview mode. Preview dialing will be covered further in the Advanced Features Section.

HELP TRANSFER

The predictive dialer should permit the transfer of a live answer call (voice and screen data) to a specialist/help agent.

AGENT MONITORING

The predictive dialer should permit supervisory personnel to observe agent call handling. This should include hearing both sides of the conversation as well as observing the agent's data screen activity. But before initiating agent monitoring, one should thoroughly investigate the legality involved with this procedure.

TELEPHONY CONNECTION

The predictive dialer's inbound/outbound telephone lines and agent's telephone sets should connect (if desired) to the existing PBX/ACD. Telephone lines used for outbound dialing should optionally connect directly to the telephone service provider. You should have a choice of analog or digital connections.

CONNECTIVITY ENVIRONMENT

The predictive dialer should adapt to the existing mainframe(s), server(s), LAN and/or WAN connectivity environment. A company should not be expected to make logical changes in these systems to accommodate a dialer. At times LAN network segments need to be isolated and/or connected via routers due to existing high traffic on the LAN. These changes are strictly traffic related and would normally be recommended by network management personnel.

REPORTING

The predictive dialer should offer a set of standard real-time and historical reports. A more detailed discussion of reports is found in the Advanced Feature Section.

OPERATION

Two important operational features of the predictive dialer:

1. The system administrator should be able, during operation, to manually change the dialer's behavior for the application and the agent.

2. The normal mode of dialer operation should be unattended. It should not require a system administrator to constantly monitor conditions and change behavior settings to assure proper operation.

TELEPHONE AGENT WORKSTATIONS

Equally as important as the previous 16 items: predictive dialers work best with intelligent workstations, usually some type of PCs. Companies may have a corporate standard for type and system software. Unless the workstation and software are very ancient (we won't try to define ancient), the predictive dialer should be able to use the chosen standard.

Probably 95% of our installations utilize IBM or IBM-compatible PCs. The same 95% use DOS, Windows or OS/2 as the workstation operating software.

When deciding which predictive dialing system to purchase, intelligent workstation capabilities are extremely important. The dialer vendor should easily answer: "What kind of magic can you do with workstations?" Software provided with the predictive dialer should automate many of the complex manipulations agents do while using the legacy mainframe applications.

We've shown how predictive dialers increase the productive time of telephone agents. Good workstation software implementation adds significant productivity by dramatically decreasing the amount of time spent on each contact! (See Chapter 7, Single System Image View -- Workstation Software.)

ADVANCED FEATURES

We've looked at the basic features offered by predictive dialing systems. Now we need to examine the advanced features, which vary from vendor to vendor.

INBOUND CALL HANDLING

The predictive dialer should permit the same telephone agents to process both inbound and outbound calls at any time. It should be customizable to automatically answer inbound calls with your recorded messages, music, etc.

On T1 telephony connections, the DNIS/ANI data should be captured and used by the predictive dialer to notify the agent why the person is calling (DNIS) and from what telephone number the call originated (ANI). If your databases or host applications permit access by the caller's phone number, the predictive dialer software should provide automatic retrieval and workstation presentation of the caller's data record.

In addition, through the built-in automated attendant, the predictive dialer should be able to ask the person to key in their account number. In this mode, the dialer software should use this information to access and present the appropriate record on the agent's screen.

One last feature is the ability for the agent to indicate to the predictive dialer not to call this person again today. This feature is similar to the function that automatically redials a person at a specified time only if that person has not returned a requested call.

PACING LOGIC

Because all calling campaigns have different characteristics, one pacing method won't fit all applications. Multiple pacing methods are needed to establish the desired behavior. The predictive dialer must be able to handle very short calls followed by very long ones and still maintain a low nuisance call percentage rate.

VIRTUAL AGENT

What if the call center has 100 telephone agents and each has an outbound component of their work. Let's assume that component is 50%.

With proper scheduling, it appears that the call center manager would only need a predictive dialer that could handle 50 agents at a time!

If dialer agent positions are physically fixed, the manager has three choices:

1. Equip all 100 agent workspaces with dialer capability. This could be expensive, especially if the agents are only using the dialer half of the time.

2. Set up a special dedicated work area of 50 positions. When agents want to use the dialer, they physically move from their desks to one of the dedicated dialer work areas. This, too, sounds expensive! The added space, extra workstation computers, lost time and inconvenience all make this option very impractical.

3. The best solution is for any 50 of the 100 agents to be able to log onto the dialer from their regular workplace, wherever it may be. About the only requirements are that the 100 agent workstations be on a LAN/WAN and all the telephone instruments be connected to the PBX/ACD.

This predictive dialer capability is known as Virtual Agent, because it lets any agent log onto and use the dialer from any agent position equipped with the standard hardware and workstation software the dialer requires.

CTI INTEGRATION

We are frequently asked, "Do you support Computer Telephony Integration (CTI)?" We say, "Yes, how do you plan to use CTI?" The usual answer is, "We don't know yet; we just want to make sure we buy a dialer that supports CTI."

CTI covers a lot of ground — like water! Now is this water for drinking, bathing, boating or swimming? When purchasing a predictive dialer, be precise about all possible uses, so the needed CTI capability is deliverable by that dialer.

CTI Elements are Beginning to Merge

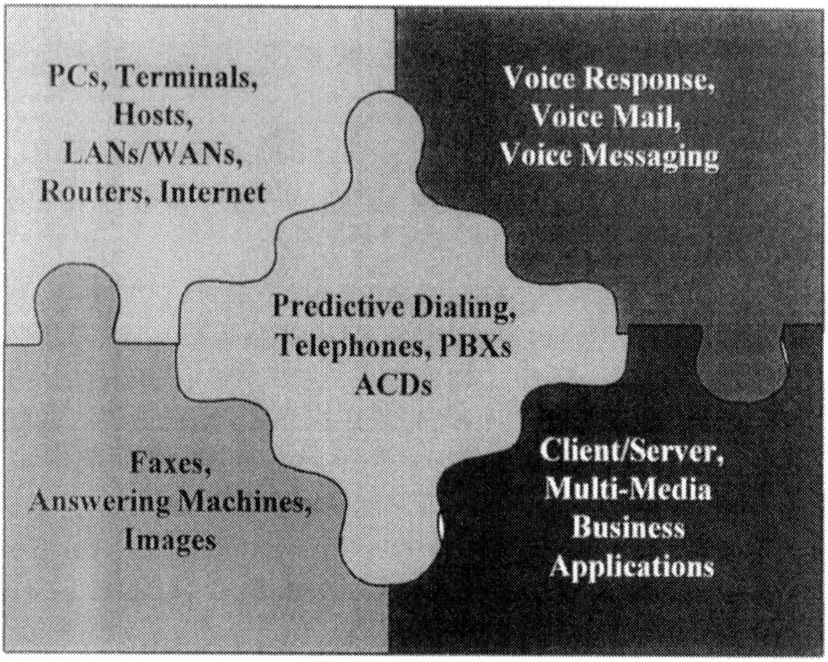

Figure 15
CTI is More Than Computer Telephony Integration

RECORD KEEPING

Due diligence is desirable, even mandatory for some applications. The predictive dialer selected must be able to automatically upload (in real-time or batch mode) to your host computers, a complete record of every call attempt made, regardless of its outcome. The record should be date and time stamped and include:

- Number called
- Account number
- Agent ID (if handled by an agent)
- Anything else germane to your business

The predictive dialer should also allow the system's administrator to audit long distance billing if call length is included, keep a record of each call attempt for due diligence reports, and even help develop a profile on each client's best/worst times to call.

CUSTOM REPORTING

Like the saying, "No two people (even with the same job) are alike," no two predictive dialers are alike. Especially when it comes to reporting. A predictive dialer should provide the system administrator open access to the data and serve as a tool for creating customized reports. The dialer's solution should allow you to use virtually any off-the-shelf reporting package (Microsoft's Access, Lotus 123, Crystal Reporting, etc.) that uses industry standards like ODBC. But before tackling custom reporting, we suggest someone from your call center attend training on how to do this. While customizing reports is not a full-time job, it is important to know how to do it correctly.

THE WORKSTATIONS AGAIN

Many of our customers have grown tired of outdated, proprietary, character-based representations of legacy mainframe data. These terminal screens are usually cluttered, hard to read, poorly organized for the telephone agents use and require manually jumping from screen one to screen six, back to screen three, etc.

In addition, users need to simultaneously display information from more than one mainframe and/or database. To this end, some have developed (at considerable expense) computer program(s) that run in the workstation. These hide all the legacy data, bring together customer information from multiple sources and present this in a manner consistent with the agent's ease-of-use requirements. These workstation application programs further insulate agents from all the usually different keystroke rules each legacy system demands.

With these programs, the workstation makes use of graphics when appropriate, along with common definitions of keys and icons for all hosts/legacy system applications. This really improves agent effectiveness and creates opportunities for agents to satisfy all customer needs during one contact. Better informed agents are empowered. Empowered agents deliver better Customer Care. Now some predictive dialer suppliers offer a comprehensive yet intuitive, easy to use development platform to do this for a call center and an entire enterprise.

WORK FROM HOME

Some predictive dialer suppliers are offering a work-from-home capability for part-time, disabled or some full-time agents. Working from home should operate as though the agent was at your call center. There should still be seamless access to all information and telephone services. This work-from-home option is rapidly growing.

NUISANCE CALL CONTROL

We talked about this in many prior sections. When comparing predictive dialers, one needs to fully understand how each dialer supplier accomplishes nuisance call control. A vendor's answer, "It is proprietary," is unacceptable. Nuisance calls will happen. So the key is how to correct the condition before it creates a nuisance call. Is the method offered by this supplier acceptable to you?

CALL CONTROL LOGISTICS

Let's assume it's 5 p.m. sharp. The predictive dialer has been dialing for a while. At this exact time, the dialer application has:

- 200 no-answer calls to retry
- 20 busy-signal calls to retry
- 5 previously answered calls, each scheduled for a 5 p.m. retry
- 500 people who have not yet been dialed

A cursory review would indicate one way the dialer can tackle the calls is in order. Logic would dictate that the five appointed-time retries are the most important, with the 20 busy-signal retries next (you know someone is home). But then what? Well, what's more important? Should the agents retry the previous no-answer calls next or attack those not yet tried even once?

The point: the system administrator should determine the order for each application. The dialer's behavior control should be user selectable to assure that the desired priorities are executed.

Many other controls are useful:

- Alternate phone number attempt

- Alternate person attempt
- Different rules for different times of day
- A call back by another agent in a few minutes to verify the results of the original contact

SPECIAL CASE HANDLING

Let's assume that we just created a nuisance call. A person answers the phone, no agent is available and the predictive dialer disconnects the call. What now? Can we prevent the nuisance call from happening again to the same person? How should we handle it? Your predictive dialer should permit these options:

- Not calling again today.
- Calling back in 30-60 minutes but only in preview mode. This assures that a hang-up won't happen again.
- Calling back in preview mode with the next available agent. The agent can then apologize for the disconnected call and proceed with business.
- Each call center will vary in how it may handle nuisance calls or call backs.

Other significant special case call handling options you may need:

- Pass on to agents all calls that result in the 3-tone signal (SIT tones) that indicates disconnected or changed numbers but not press other conditions.
- Allow the predictive dialer to filter out all special tones and just report on these.
- Set different priorities for different time zones.
- Set different priorities for all call dispositions (not called, busy signal, no answer, different time zones, answering machines, etc.) for each campaign.
- Set priority of inbound call routing to various groups.

We could go on and on with option after option, but the question is: What does management control mean? The more control, the greater your opportunity to put together business processes to best meet your corporate goals.

WHAT TO AVOID

One would think common sense would be the ruling criteria when choosing the right predictive dialer. And to a great extent it is. For example, a chosen predictive dialer vendor's history of financial success indicates the products and services they offer appealed to those who purchased their system in the past. A sizable commitment of a firm's income to research and development and to service are good indicators the company is keeping its product current and supporting its existing customers.

Other things, however, might not be quite so obvious. Here are a few of the most important considerations:

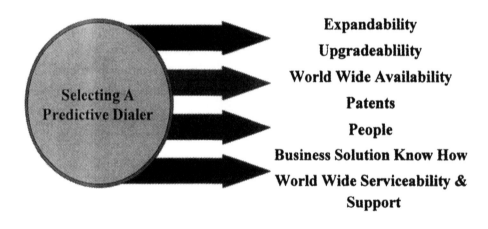

Selecting A Predictive Dialer

Expandability

Upgradeablility

World Wide Availability

Patents

People

Business Solution Know How

World Wide Serviceability & Support

Figure 16
Predictive Dialer Essentials

EXPANDABILITY/FLEXIBILITY

When a predictive dialer is purchased, it becomes equipped with a certain number of agent positions and a number of telephone trunks. Eventually most predictive dialer users want to increase both these components because of the success they have with the system.

Will more seats or trunks be added? Maybe. Maybe not! Can the model under consideration be expanded? A good rule of thumb is to purchase a model that can be expanded (as opposed to replaced, whole or in part) to twice its original size. For example, if the system chosen starts with 30 agent positions and 50 telephone trunks, then it should be expandable to 60 agent positions and 100 telephone lines and on the same existing platform without sacrificing flexibility, speed and functionality. Expansions should not require additional:

- LANs (a single LAN should support all agent workstations, supervisory workstations, printers, etc.)
- Switches (a single switching platform should support hundreds of agents on a nonblocking basis)
- Multiple servers for adding only few agent positions

System reliability and ease of support totally depends on the number of components required to meet the call center's mission. The more dedicated/proprietary components, the higher the failure rate and cost of operation.

UPGRADEABILITY

Over the years, the capabilities of predictive dialers have increased significantly. New and improved features and functions increase opportunities to better control the ability of the dialer to execute business processes. These new features and functions require software changes to the system. Frequently they also require hardware changes and/or additions.

One factor to check when selecting a predictive dialer vendor is to consider their historical philosophy (over a period of 6-8 years) for upgrading their existing customers to the most current solution. Does the upgrade require a complete replacement (often referred to as a fork lift upgrade)? Or, has the supplier provided a method for existing customers to take advantage of the new capabilities at an incremental cost? Upgrading may require adding or replacing some hardware, but this is certainly preferable to the cost of a complete major component replacement just to stay current. Look at the technological changes we're all experiencing. Things today are changing faster in 18 months than previously in 10 years.

Any investment in a predictive dialing system is more likely to be protected by a supplier who has a several year history of adopting and demonstrating a customer migration approach. Avoid those who require major replacements.

WORLDWIDE AVAILABILITY AND SUPPORT

Another factor that must be considered in any predictive dialer selection is worldwide product coverage. If the company buying the dialer plans to be a multinational company, then it needs to know the dialer supplier has demonstrated support for its products in countries outside North America. Words are cheap here. A vendor saying, "We plan to..." is not satisfactory. Marketing products, especially predictive dialers, in foreign countries requires a significant investment to provide local service/support and ongoing training, in addition to specific approvals. Has the chosen supplier already made these investments? What investments has the supplier made toward assuring your continued success? What is the ratio of their support personnel to the number of customers they support? What percentage of their sales revenue is being spent on support services? What infrastructure exists to ensure that you among hundreds or even tens of thousands of customers (that supplier may support) are treated as NUMBER ONE? What does the supplier's support team look like? What responsibilities does the supplier assume? (More on this in Chapter 6.)

BUSINESS SOLUTION KNOW-HOW

Features versus solutions is the question. We have all been exposed to sales approaches where the bells and whistles are presented with overwhelming gusto. Most of the time, it's difficult to relate these (in their raw form) to business solutions.

Does the team of people representing the predictive dialer supplier understand enough about your business to help you achieve your objectives? To use our car example: if one supplier claims his automobile's air conditioner will hold a 0° inside temperature in the middle of the 120° Arizona desert, and the other supplier claims only 65°, does it really make any difference? Only if the car is going to be used for a deep freeze. Predictive dialer features are great, but solutions to business problems are probably more important.

So back to the original question: Will the predictive dialer's team deliver solutions to business problems or are they just selling a 0° freezer? Features usually sound very impressive. Unless they are relevant to the business solutions presented, it's up to the buyer to figure out how to use them. How important is it to have a car that can go 225 miles per hour on roads with a 65-miles-per-hour speed limit?

INNOVATIONS AND PATENTS

Just a short comment. Are any of the key patents held by the chosen supplier likely to give the system user a competitive advantage? Conversely, is the buyer protected from patent infringement by the chosen supplier? Is the buyer going to win more business by continually deploying these innovations ahead of his competitors?

PEOPLE

Finally, remember that every person on the predictive dialer supplier's team -- from the salesperson to the installer to the receptionist — represents the values and knowledge this company offers. The team comes with the dialer. Before deciding on which company to partner with, ask: "Can this team help me design and deploy a good, long-lasting and flexible business solution?"

CHAPTER SUMMARY

In this chapter we've described some of the major points to consider when selecting a predictive dialer. We submit to one additional truism. Dialer technology is rapidly progressing.

If one buys only features, then in six months, another dialer supplier will have more features. And then another feature, and another, etc. So what's the answer? Select a solid, client/server predictive dialer platform on which business solutions can be built. Select a business solution you and the supplier have jointly developed. A poorly designed solution using all the best features is still a poor solution. A well designed solution combined with a lasting partnership generates a strong bottom line.

Chapter 6

How to Be Successful in Implementing a Predictive Dialer

As we struggled to finalize details in the late spring of 1990, one customer was preparing for the installation of the predictive dialer. All the paperwork was signed and the system integration plan was complete. The customer faxed information on the readiness for both personnel and the site. They let us know that they were awaiting the final system shipment.

On our side, we completed and reviewed all of the final details ensuring the equipment had been staged, the software with customized features and functions had been tested and integrated and our project management and field engineering team was ready to commence the installation.

As we arrived at the customer site we were asked to wait so the customer could physically locate the equipment. That wasn't that unusual because frequently when we arrive the customer needs to locate the equipment since it does not always get delivered to the right floor.

As it turned out we were told the equipment was on the 12th floor. Our people took the elevator and arrived at the 12th floor. As they exited the elevator, they started to notice a bright light coming down the corridor, plus a loud sound that became louder as they approached the call center door. Suddenly, as they opened the doors, they realized why they could see the bright lights and hear the loud sound. The sound was heavy rain coming down. Yes, the equipment was there, the boxes were standing against the wall. Looking further out, and to their amazement, they noticed that there was only a partial roof and construction was still going on. Some other doors on the floor weren't completed so that's why light came through into the corridors.

It took us a few days to get it all ready for installation. There was no electricity, no desks, no working spaces, and the rain was pouring down onto a concrete floor. We sure had a problem.

Eventually, the problem was resolved. The equipment was pulled out of harms' way, the building was completed and the predictive dialing system was installed and up and running successfully.

It's critical to the success of a project to communicate, assign roles, develop expectations on both the customer side and the vendor side. To date we have never faced this type of situation again. System integration plans are developed early on and all members of the vendor/customer team know their roles and what's expected.

In this chapter we discuss people, issues and steps involved in implementing a predictive dialer into an organization. The ultimate goal of any implementation is to complete it quickly and successfully. This requires a real partnership with the chosen predictive dialing supplier and a project team dedicated to success.

Just having the right people is not enough. These individuals must be supplied, even empowered, with clear objectives that detail how your management expects the project to be completed. A successful implementation will also require:

- Detailed planning
- Effective communication among team members
- Individuals accountable for not only their responsibilities, but also the overall project
- A careful examination of the technologies

SUPPLIER/CUSTOMER PARTNERSHIP

The early stages of any partnership are not unlike a marriage immediately following the honeymoon. There is a period of time it takes time for the two partners to get to know each other's needs, desires and expectations. The key to forming a successful partnership with a predictive dialing supplier is to document all your requirements and expectations. This prevents later problems due to misunderstandings, changes in personnel or issues that arise during the project implementation.

If this is your first predictive dialer purchase, you may not be able to provide the supplier with the technical requirements necessary to fully implement your chosen solution. Make sure the supplier knows exactly what you expect from the dialer after it is installed — performance expectations, what systems it needs to integrate with (PBX/ACD, mainframe, databases, servers, LAN/WAN, IVR/voice mail, etc.), what the agent interfaces should look like and what kinds of equipment are currently being used.

The supplier should then be able to develop an implementation plan that includes detailed technical requirements for the system, the system's performance specifications including reporting, system architecture and agent interfaces.

To ensure a smooth implementation, the supplier also needs to define who will provide what equipment and when, all time-lines, plus drawings and physical layouts. This clearly documented information establishes the technical and operational baseline.

All contracts such as warranty, maintenance type and level of service should be well documented. Misunderstandings on these issues, which primarily appear after a predictive dialing system is implemented, can be avoided by clearly defining and documenting all agreements prior to the purchase.

PROJECT TEAM

A successful project implementation requires a team of individuals from both sides dedicated to meeting the established objectives. They must have the responsibility and authority to carry out the plan. Although the term authority is used in various ways by management experts, the standard definition is "legal or rightful power, a right to command or to act."

In a project management environment, authority is the power to command others. It is the basis for all responsibility. On the other hand, it is the obligation of a subordinate to perform all assigned duties. In a typical implementation project, the project manager has the overall authority and responsibility for the project's successful completion.

Today's highly technical, complex, computer-telephony-based systems demand an expertise no single individual typically possesses. This necessitates that the predictive dialing supplier bring together a team of people with diverse technical and functional skills to accomplish the task of properly analyzing, defining, designing and deploying a complex, sophisticated telecommunications solution.

THE CUSTOMER PROJECT TEAM

A predictive dialer installation is unique in its far-reaching effects on your operation and infrastructure. Properly planned and implemented, it will be successful and not be disruptive. There will, however, be an impact. There is a definite need for an empowered project manager from the customer's organization who has an overall understanding of the business objectives, applications to be processed by the predictive dialer and the agreed upon operational characteristics of the predictive dialer in the customer's environment.

Ideally this project manager should be involved from the beginning of the evaluation and study, both of which occur well before the selection of a predictive dialer supplier. The project manager should be involved in all dialogue between the potential suppliers and your operational and technical management. This is the phase where your overall requirements and expectations should be finalized. By participating in these phases prior to supplier selection, your project manager will have a good understanding of what you expect and how the chosen predictive dialer supplier will deliver and meet the agreed upon results. After supplier selection, your project manager must be the boss.

Your predictive dialer typically interfaces with almost all your data processing and telephony systems. Specifically, agent workstations, LANs and maybe WANs, mainframe computers, database or network servers, PBX/ACD, and possibly IVR systems. Additionally, telephone agents and their supervisors and managers will experience changes to their operating procedures and environment. None of this should be cause for any concern. All these facets must be considered and covered in the project plan. It is too much to expect that the project manager be an expert in all these areas. So your project team must include people who have the necessary skills.

Now that you have your project team, it is important that you document and convey to your chosen supplier any and all corporate standards.

Examples might be:

- Do you have a standard workstation platform, a software set in the workstation, a standard network interface software/hardware?
- What are the details of your LAN topology, routers, network operating systems and releases (include also network traffic measurements)?
- If your PBX/ACD is involved, what functional details do you want to preserve? If the predictive dialer is to connect to or

through such a switch, what is your preferred connection method?
- What protocols are used to access/communicate with your mainframes, databases and application servers?
- What changes to any of these are contemplated and when?

This is not meant to be a comprehensive list, only a sample.

From the operational side, the following must be defined:
- Telephony agent workstation navigation and operation
- Supervisory and management roles and responsibilities (who does what?)
- Supplier's policies, procedures and cost structures for services
- Calling list/calling strategy and application processes

Again, not a complete list, only examples.

These areas must be documented, reviewed and refined in conjunction with your chosen supplier's project team. The importance of formal documentation cannot be emphasized too strongly.

Obviously, the logistics effort must be planned:
- Where are the various components to be installed? Getting the site ready.
- Are the required electrical, data and voice terminals for connection to the predictive dialer installed in accordance to written specifications and the installation time-line plan?
- Is there a phased-in plan that includes training for supervisors, administrators, agents and project managers, plus on-site testing?
- Have you arranged off-hours access to your facility by your supplier installation team?

Installation of a predictive dialer is a team effort and will be smooth if the customer and supplier project teams effectively communicate and execute their respective parts of the joint implementation plan. The objective is to make sure each team understands each other's roles and that they jointly eliminate "I forgot to tell you..." situations.

SUPPLIER PROJECT TEAM

The typical predictive dialing supplier's project team will be comprised of a project manager, systems engineer, configuration engineer, field services engineer and the salesperson who sold the predictive dialer. In some cases, one person may fill multiple roles. Though different roles are played by individuals, the team goal is a successful implementation.

SUPPLIER PROJECT MANAGER

Similar to the customer's project manager, the supplier's project manager will direct the team's implementation effort. The supplier's project manager should have broad authority over all aspects of the entire project. Once the contract is signed, the project manager manages and tracks details involved with specific installations, upgrades or customer moves through implementation, training, acceptance and final transition to the supplier's Customer Care Center for ongoing support.

The project manager should have the proper tools not only to manage the project, but also to provide implementation status updates to all concerned parties. A number of software packages are currently available to aid project management in establishing time-lines, deliverables, responsibilities, etc. The software selected depends on the specific needs and complexities of the project, and as such is an essential tool for successful management.

Companies generally implement a predictive dialer as part of a larger change in their internal systems. Both project teams (customer's and supplier's), simply referred to as the TEAM, must not only manage the implementation of the predictive dialer, they must ensure that this implementation supports the corporate mission.

SYSTEMS ENGINEER

The systems engineer is responsible for the technical aspects of the project and coordinates with the customer and internal departments to ensure that all customer objectives related to installation, integration, training and repairs are met.

Working closely with the MIS staff and call center personnel, the systems engineer analyzes and then documents the environment in which the predictive dialing system will be deployed (e.g. host, type of emulation, Ethernet or token ring, LAN/WAN, T1/ISDN networks, PBX/ACD type, etc.).

Accurate documentation of this information is critical to the success of the project. It forms the basis of the work to be performed by the TEAM. It also impacts the schedule and delineates what must be done to prepare the site for system installation and implementation.

CONFIGURATION ENGINEER

The configuration engineer is responsible for configuring and/or customizing the unique application software required to ensure the system operates as desired.

The configuration work depends on the systems engineer clearly defining what information and what type of presentation screens the customer wants. The automation of transaction processing, CTI applications, scripting, database object definitions, host uploads/downloads, etc., are the responsibilities of the configuration engineer. The configuration engineer may also work with the customer and end-users if questions arise during implementation.

FIELD SERVICES ENGINEER

The field services engineer is responsible for the actual installation and implementation of the predictive dialing system. This person may also be responsible for training the agents and supervisors who will be working on the system. Prior to being shipped to the customer, a predictive dialing system, with its major unique components, should be staged by the field service organization. This helps determine if each major component works as intended and that the entire system performs to specifications.

SALES

The salesperson starts the solution process by working with the customer to identify objectives and help structure a solution. The salesperson is responsible for the sales cycle from initial contact to contract signing and production cut-over, ensuring that the solution sold meets the customer's needs and expectations. Specific responsibilities include:

- Cultivating new customers as well as new opportunities with existing customers

- Analyzing customer business needs
- Educating customers on products, services and policies
- Developing and proposing appropriate solutions
- Following up with information to support customer requirements
- Ongoing relationship building

The salesperson is also a key contact in helping resolve issues that may not be answered by the project manager.

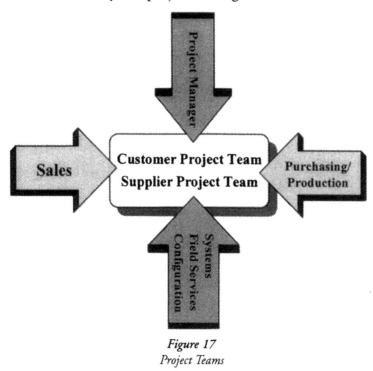

Figure 17
Project Teams

OTHER FUNCTIONAL AREAS

There may be times when support for the project is needed from other functional areas such as:

- Consulting
- Software or hardware engineering
- Training

It is the project manager's responsibility to identify the resources needed and work with the respective representatives or suppliers to assure the project stays on schedule.

PURCHASING/PRODUCTION

Purchasing/Production functions include buying and assembling specific hardware and software components required by the predictive dialing solution. If unique parts are required they are ordered by purchasing in time to support installation and implementation.

Production builds and tests the predictive dialing system to the specifications produced by systems engineering and the TEAM. Production uses well-defined processes and procedures to assemble the system's major components and customized applications, then provide a comprehensive test and analysis of the entire system.

The project manager oversees all scheduled production and purchasing due dates to determine any variances and institute corrective action as needed.

ACCOUNTABILITY

The individual members of the project team are responsible for their particular areas of expertise. Each member of the TEAM accepts accountability for their area. When issues arise, the appropriate member resolves them expediently to prevent or minimize any schedule or performance impact. Management remains accountable for issues that may arise at their level, expediting resolutions to minimize schedule changes.

ROLE OF PLANNING IN PROJECT MANAGEMENT

A wise man once said, "Proper planning prevents poor performance." This is never more true than in preparing for a successful predictive dialing system implementation.

Planning involves choosing from alternative courses of action to determine what to do, how to do it, who will do it and when it will be done. Planning takes us from where we are to where we want to be. It is the most basic management function and is imperative to the project management team. A key step to establishing a good plan is to set objectives.

Each objective must be clearly defined, assigned to a responsible individual and have a completion date. Together, this set of unique objectives constitutes the plan. However, a plan by itself cannot ensure success.

The project TEAM is responsible for periodically reviewing their progress against the plan and reporting how implementation is proceeding. The TEAM must take any necessary corrective action if there is a deviation from the project plan.

ESCALATION

There may be issues during implementation that cannot be immediately resolved at the project management level. It's important that both customer and predictive dialing supplier management establish and use an escalation policy to deal with such matters.

The policy begins at the lowest level of authority and works its way up in the respective company's management chain until the issue is resolved. The intent is to resolve the problem at the lowest possible management level prior to escalating it. This can be an effective tool in problem resolution because no one wants the boss to know there's a problem they can't resolve. Timing is everything. Resolution must be implemented quickly so it will not adversely affect the success of the project.

COMMUNICATION

In today's environment, we are inundated with methods of communication: telephone, voice mail, E-mail, pagers, cellular phones, fax machines — just to name a few. Yet communicating is one of the most difficult things people do. In an implementation, communication between all affected parties is vital! Every person in a project environment is responsible for good communication. Each communication should have a purpose and that purpose should be to influence action in the direction of the project.

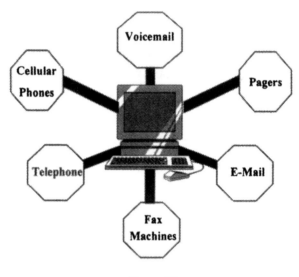

Figure 18
Communications

Successful communication requires an information system. It should be an open loop that allows the project manager to specify what inputs are needed from the project team. The information flow system should have four basic elements:

- Collecting
- Processing
- Comparing
- Selecting

The project manager on each side should summarize the results for all involved. All affected functions or individuals should be kept informed of progress, problems and resolutions.

WALK BEFORE YOU RUN

With the new technologies available, companies are often tempted to completely revamp and implement a system with all the attendant bells and whistles. There are a number of questions to answer before undertaking a major system implementation.

What are the hidden costs of such an undertaking? For example, what will be the costs of training all affected individuals on the new application software and hardware? If agents are primarily used to working in an ASCII terminal or DOS based system, then the OS/2, MAC or Windows applications dealing with graphical user interfaces may have not only training issues, but also a significant learning curve as agents become proficient.

What impact could the loss of productivity during the learning curve have on a business?

To determine if it is a wise business decision, it is essential to look at the cost of meeting these needs versus the increased revenue and other benefits you expect.

Another factor to consider is the sophistication of the staff who will manage the predictive dialing system. The system administrator should be someone with a business and computer background. They should be familiar with the type of environment being installed, e.g. UNIX, Windows, OS/2 operating environment, LAN/WAN client/server based applications, etc.

Depending on the application complexity and system size, there are other cost considerations: hiring, training, turnover and a multitude of other issues.

CHAPTER SUMMARY

A successful predictive dialing system implementation process is multifaceted and requires detailed planning and effective management. Project managers from each company must communicate effectively at all levels, and inspire a commitment from all members of the TEAM toward immediate problem resolution. It is indeed a partnership —TEAMwork, and if properly managed, can be very rewarding and profitable.

SELECTED REFERENCES

Koontz and O'Donnell, *Principles of Management: An Analysis of Managerial Functions,*

Chap. 4, New York, McGraw-Hill Book Company, 1968

Cleland and King, *Systems Analysis and Project Management,* Chap. 10, New York, McGraw-Hill Book Company, 1968

Chapter 7

Tomorrow's Customer Care Focused Business

BUSINESS IN THE DIGITAL AGE

The dawn of the digital age brings many challenges and opportunities to businesses. Businesses able to adapt to changing consumer behavior by applying new technologies will succeed. Those tied to outdated customer models and obsolete technologies will fall behind. In this chapter we will look at how technology is changing customer needs and behaviors, and how businesses can leverage these new technologies to establish lifetime links with customers, a focus we coined in 1993 as Customer Care™.

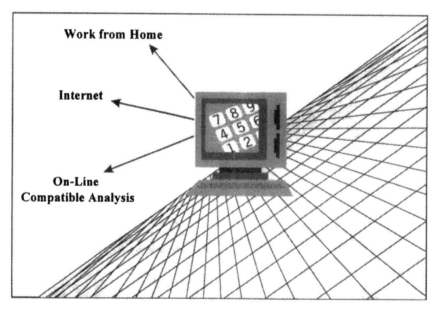

Figure 19
The Digital Age

THE DIGITAL AGE

The world is rapidly changing because the standard by which information is exchanged has virtually moved from analog to digital. This change presents many opportunities for consumers and businesses.

Consumers will:

* Buy products and services, and get the best deals through online compatible analysis

- Work from their homes using voice, video conferencing, teleconferencing, fax machines, etc., on their PC's, running group collaboration and communications software creating a virtual office
- Access information on just about any subject, from tens of thousands of sources located across the globe via Internet (and its successor) and other global networks.
- Have a higher quality of life because of better understanding and more choices

Businesses will:

- Have direct global access to consumers and markets
- Have increased alternate channels for information and product delivery (historical ways via cable TV, mail order, newspaper, etc., and new ways via Internet, interactive TV, interactive radio, satellite, power companies, etc.)
- Be able to dramatically reduce the costs of advertisements, information delivery, communications with customers, vendors and suppliers, human resources overhead, plus much more due to the Information Highway and/or Information Services Supermarket where people and nonhuman resources can globally collaborate.

This at-home access to technology will bring new information, remove geographical barriers and enable consumers greater selection in who they do business with, and how their business is transacted. They will be able to more readily analyze products and services and quickly choose what best fits their needs.

As companies begin to exploit the capabilities of these new digital technologies, consumers' choices and, subsequently their expectations, will expand. For example, a cable company may soon expand its current offerings to include full interactive services. The cable subscriber who used to choose from 50 channels will soon select from more than 500. In addition, the cable company may offer its customers local telephone service, Internet access, video conferencing, shopping services, and more.

With more choices available, consumers will become more demanding of the companies with whom they do business. They will expect to be able to customize services they purchase. In addition, consumers will be able to complete more research before making any purchase to identify the product and service that best fits their needs.

An important consideration for businesses will be that consumers are no longer restricted to choices in their immediate geographic location. A variety of communication media and methods will give consumers virtually free and immediate connection to global information. They can as easily choose to do business with a company across the country (or even across continents), as they can with a company across the street.

WHY DIGITAL TRANSMISSION IS IMPORTANT

Before we continue, we need to talk a little more about how the advent of the digital age is enabling the creation of systems that can support the Customer Care business model.

Continued quantum leaps in computers and systems capable of making intelligent decisions based on the content of information will become pervasive. The computer can be taught to respond to an E-mail, a fax or even a customer's voice request. It may complete the task on its own, or automatically route the information to the person best equipped to do so.

Say a customer sends a letter to your business asking for more information on a product. Your computer can:

- Read the text using optical character recognition technology
- Determine what product the customer is interested in
- Automatically send a personalized information packet back to the customer via Internet, E-mail, fax, etc.
- Schedule a future follow-up call by the predictive dialer

The computer could accomplish this without any human interaction, saving your company time and money. Your salespeople could focus on closing as opposed to supporting business.

THE CUSTOMER CARE FOCUS

If you think all this increased competition creates concerns for businesses, you're only half right. You see, while these technologies will dramatically change customer behaviors and expectations, they will also provide businesses with next generation tools to help them establish lifetime links with their customers. From a predictive dialer point of view, look at the effect it has on the Business Life Cycle and Customer Care:

- New customers are acquired through targeted outbound calling campaigns.
- After-sale follow-up campaigns are launched to assure the highest level of customer satisfaction.
- Customer service campaigns insure prompt response and satisfactory results to any customer-generated issues.
- New telemarketing/telesales campaigns are directed at existing satisfied customers, selling additional products and services.
- Customers as early as 3 to 7 days delinquent on their payments are automatically included in the outbound friendly reminder campaigns.
- Customers with accounts reaching 30-45 days overdue are automatically included in outbound debt collection campaigns.
- Outbound campaigns periodically inform customers of new products, company news and significant events.
- Proactive surveys give customers the opportunity to share views on products/services and their vision for the future.

As outlined above, even the basic predictive campaigns if implemented as part of a Business Life Cycle can provide a competitive edge. Businesses must now begin to utilize advanced technologies to deliver more, and therefore have the opportunity to forge lifetime links with customers. This idea of focusing all of your daily business activities on serving your customer's needs is what Customer Care is all about.

Figure 20
Focus on Customer Care

Businesses must provide customers with outstanding products and services by knowing and delivering exactly what their customers want. This requires that the business maintain an ongoing relationship with this customer, where every major step in the business process brings the customer closer to the business. To accomplish this, every employee must focus on strengthening this relationship. Every customer interaction must have the goal to make that customer a Customer for Life.

In the world of rising costs, predictive dialers, and the systems with which they interact, will become an integral part and strategic link between tomorrow's profitable businesses and the savvy customers they serve.

Today's consumers have less patience and spend less time on product quality and service issues. With all the choices available to customers, businesses must do it right the first time because they may never be given a second chance.

PREDICTIVE DIALING'S ROLE IN CUSTOMER CARE

While we live in an age where it seems new communication systems are being unveiled almost daily, most people still prefer to communicate over the telephone. More effectively than any other communication medium, the phone provides real-time dialogue.

Telephone media usage (wire line or wireless) continues to be the fastest growing sector in telecommunications. Current projections estimate major worldwide expansion with revenue spending that will exceed $1.0 trillion by the year 2001.

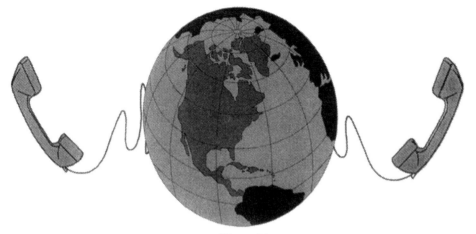

Figure 21
Telephone Media Usage within Telecommunications

So while the world enjoys increased access to information from digital sources, the most effective way for your company to reach customers is still to phone them. As we've discussed in previous chapters, no other communication medium offers your company the opportunity to make such a personal connection with your customer. This individual personal attention, this focus on each customer's needs and preferences, is the core of our Customer Care paradigm.

Predictive dialing systems make all this possible at a relatively low cost. They enable your employees to connect with your customers as needed, and provide the information to develop a Customer Care environment. Without the cost-saving help of predictive dialers, businesses will not be able to afford the critical telephone follow-up that can make the huge costs of printed media/mail really payoff. Using predictive dialers, businesses become empowered with the capabilities they need to consistently communicate a Customer Care philosophy.

HOW BUSINESSES WILL CHANGE

We've talked about how consumer behavior is changing as a result of the digital age, and how businesses must adapt and follow a Customer Care model where the customer is King. This may require many changes in business policies and processes.

Companies must reexamine how they do business and formulate a plan that will transform them into a Customer Care-oriented enterprise. To accomplish this, the business and call center management must:

1. **Understand the current call center environment (where applicable)**

- Transacting with existing or prospective customers (telemarketing, general surveys)

- Call processing — heavy inbound or heavy outbound or a mixture of heavy inbound and some outbound or heavy outbound and some inbound

- Multicampaign environment managing multiproducts and or services

- Transaction oriented environment with complimentary hosts, databases and server-based applications

- Real-time monitoring control and reporting on the call center's mission versus actual results

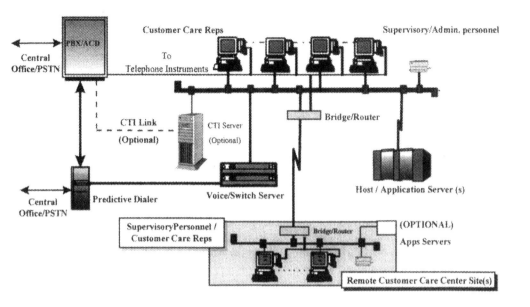

Figure 22
Predictive Dialer Based Call Center Environment

2. **Define a Clear Mission for Customer Care Center Enterprise** and develop a clear and concise set of business objectives.

3. **Adopt the Philosophy: Customer is King** and combine it with measurable business objectives, delineating every step to accomplish these objectives.

   The Customer Care business process becomes customer centric where every major step relates to the customer, thus ensuring total satisfaction. Every employee becomes a Customer Care representative and is given the authority to executive clear goals to accomplish the mission.

4. **Identify and Inventory Current Methods and Resources Used to Run the Existing Call Center (where applicable)**

   In existing call centers, how customers or prospects communicate with the centers differs widely.

   In most cases, however, telephones are the predominant medium and are used to process incoming/outgoing voice calls, as well as E-mail, faxes, videos, etc.

In some cases, agents/customer care representatives provide telephone services from their homes (work-from-home), or from satellite offices.

Regardless of the situation, before a call center can be transposed into a Customer Care Center with full automation (PBX/ACDs, predictive dialing, Dynamic Inbound/Outbound™, IVR, voice mail, workflow, etc.), methods and resources currently deployed will need adjustments, changes or even total overhauling.

5. **Review Call Processing Requirements**

Only by clearly identifying the requirements and attributes for the current call processing mission, can Customer Care-oriented automation be introduced to the call center. By introducing automation such as predictive dialing, Dynamic Inbound/ Outbound call processing, etc., both a higher customer-satisfaction level and a dramatic reduction in operational costs can be attained. Future (in 2 to 4 years) call processing requirements and potential growth should also be identified at this point.

Look at current call processing attributes:

- Account preview time -- before dialing
- Right party talk time and wrap-up time
- Wrong party talk time and wrap-up time
- Time to process Busy Signal/Line Engaged call
- Time to process No Answer call (3 to 5 rings)
- Time to process telephone company recordings
- Time to process answering machines
- Time to process calls into voice mail, fax, modem
- Time for other conditions such as no ring back, fast busy (no network)
- Average·time awaiting incoming calls, etc.

Examine the call center's operations:

- Hours of operation each day for inbound campaign, outbound campaign
- Days of operation per week (5,6 or 7)
- Number of inbound campaigns
- Number of outbound campaigns
- Total number of inbound contacts processed
- Total number of outbound contacts processed

Deploy an automated Dynamic Inbound/Outbound call processing solution. Virtually all the nonproductive attributes can be eliminated allowing Customer Care representatives to concentrate nearly 100% of their time on what they do best — talk with customers.

6. Identify Key Technological Corporate Platforms

Identify current and future (in 2 to 4 years) intercommunication and application needs for both employees and customers. Then early selection of appropriate information systems, knowledge bases, inter-networking and desktop technologies can ensure a smooth transformation of a call center into a Customer Care Center.

With the emergence of multimedia and global inter-networking, selection of corporate standards supporting the future is key to the success of providing Customer Care.

In addition to employees and customers, information technology is the key asset of a corporation. Building a Customer Care Center with progressive and adaptable information systems will determine whether a business lives and prospers or dies and goes out of business.

To support advanced information systems, a high-speed/broadband networking infrastructure must be laid down. Proper category cabling, switching hub technology, deployment of multimedia-able desktops, inbound/outbound predictive dialing technologies (with or without PBX/ACD, CTI based integration), and enterprise-wide universal accessibility to desired resources must be supported. Building virtual care centers with Customer Care representatives working from home or any other place, yet fully addressing customer needs on-the-spot, requires a clear understanding of these needs, advance planning and foresight, plus identifying and deploying technologies capable today of delivering tomorrow's expectations.

7. Transform Your Call Center into a Customer Care Center

- Use outbound call processing to create proactive customer awareness through surveys, Customer Sensitivityô profiling, follow-ups, information updates, etc.
- Avoid on-hold practice for incoming and outgoing calls by deploying Dynamic Inbound/Outbound solutions.
- Provide timely and accurate response to customer needs through deployment of enterprise-wide Universal Access/Æ applications.
- Deploy a Customer Care Center environment based on individual customer needs.

8. Monitor, Control, Report and Measure Customer Care Results

To deliver continual, sustainable Customer Care, the call center operation requires virtually real-time monitoring of results against business objectives.

An automated self-monitoring operation could be established by deploying technologies built on standards (i.e. SQL/ODBC database accessibility, SNMP embedded components, TCP/IP routable networking, etc.). An enterprise-wide assurance of compliance with preestablished goals can be achieved by monitoring nominal operative conditions, deviations, alerts/alarms, etc. Additionally, to ensure the highest level of customer responsiveness and keep the major business objectives on target, Care Center's automation solution needs to provide for self adjust/self control.

For example, processing inbound and outbound calls should be dynamically self- controlled and on a call-by-call basis, so each call is promptly routed to the representative best suited to handle it. Similarly, calling-lists used for outbound campaigns should be self-adjusting to ensure the highest level of contact is maintained. Falling below preestablished thresholds should trigger the predictive dialing application to invoke other list demographics or call strategies to again assure the level of contact desired. Compiling results and assuring compliance requires both real-time reporting and reporting-on-demand. Being able to monitor and extract information from virtually all call center information components (PBX/ACD, voice mail, predictive dialing, host applications, etc.) is absolutely critical for the overall success. Deploying technologies that permit ease of information review and exchange by external applications provides an individual and enterprise-wide view of the Customer Care Center's success. User-customized reporting, scheduling and business objective evaluation can be easily obtained through deployment of communication components combined with information-embedded components (as mentioned above) and off-the- shelf report-generation desktop application software.

With a comprehensive plan and systems in place, you can begin managing your Customer Care Center. With a clear plan, focused management, and the best systems you have the best opportunity to succeed. To insure the ultimate success of your Customer Care Center, you will need to set up your delivery systems to carefully monitor and report on actual results. You then need to combine the results from various resources (i.e. ACD/PBX, IVR, host applications, predictive dialing, etc.) to assure your overall objectives are being met. This closed-loop monitoring and reporting will enable you to make adjustments that further enhance your relationship with your customer.

The bottom line is, your work is never done. Your customers can easily do business with your competitors. By maintaining an ongoing relationship with your customers, providing them with products they want and services in exactly the manner they wish, your customers are not likely to look elsewhere.

THE FUTURE OF PREDICTIVE DIALING

Previous sections of this chapter discussed how the Digital Age is changing consumer behavior, forcing businesses to become Customer Care focused. This section will examine some of the new technologies that will enable businesses to embrace the Customer Care approach.

SINGLE SYSTEM IMAGE VIEW™

Information has become a formidable corporate asset, particularly within the call center. Its completeness and accuracy is fundamental to the competitive success of an organization. But information is only as good as our ability to collect and present it.

Most large businesses use multiple computer systems (mainframes, minicomputers, servers, etc.) to store information about their business. The data in these systems is often stored in different formats. Worse yet, different application software is used to collect, access and manage this information. As you can imagine, coping with all these systems makes it difficult for telephone agents to access business information.

The amount of information necessary to support successful call center business activities can be overwhelming even when well organized. Traditional methods for accessing and presenting information require the user to switch between multiple sessions to complete a single transaction. Even most current workstation automation software requires that complex scripts be created.

To adopt a Customer Care business model, agents need to have instant and complete access to all of a customer's business information. For example, a customer is not willing to call one number to check an account balance, another to ask for assistance with a recently purchased product and a third to inquire about shipping status of a previously ordered product. Nor is that same customer willing to wait on the line while the agent accesses the first system, then a second, then finally a third. Instead, the customer expects to dial one number and speak to one agent, who can quickly and accurately address all situations.

The challenge: information required to answer the questions posed above is probably stored in three different systems and/or applications. This requires agent workstation solutions to display information supplied by a variety of different information systems on a single screen. A new breed of client application software is being developed that can do just that.

Agents won't have to worry about where information is, or what type of system is providing it. Instead they'll simply focus on the information content. The best part is that this software will be designed so applications can be built quickly and easily, without requiring advanced programming skills. This will allow agents to focus on the customer, rather than navigating through the systems that contain the information.

Our solution is Single System Image View™. Like its name, it gives agents one set of screens, or single view, that includes information from multiple sources. It also provides sophisticated navigation tools that guide agents through quick and accurate resolution of virtually any customer situation.

Using such a powerful new agent workstation solution, agents can quickly and more efficiently serve the needs of their customers, which is the basis of the Customer Care focus.

CREATING SIMPLE VIEWS OF PERTINENT INFORMATION

The temptation when working with the newest generation of agent workstation software will be to provide agents with a customer's complete history. The challenge: creating screens that give only relevant and necessary information.

This newest generation software is easily customizable to give agents only the information needed to quickly and accurately meet the customer's specific need. Creating simple views lets the agent focus on the customer, which is the basis of the Customer Care model.

APPLICATION GENERATION IN THE HANDS OF THE USER

MIRRORING BUSINESS PROCESSES AND POLICIES

One of the greatest benefits of providing the user with the ability to generate complete custom applications, is the flexibility to mold system behavior to mirror actual business processes. This union of process and policy is as unique to businesses as fingerprints are to people. Current fixed application structures require the user to make procedural adaptations that burden rather than benefit. The new breed of workstation products will let the user model the application based on business requirements.

ADAPTING TO CHANGE IN BUSINESS NEEDS

The most frustrating dilemma facing users is constant change. Mainframe centric systems have always required significant time and cost to make even the smallest changes to applications running on the host. Client/server implementations of applications give users control of their own destiny by equipping them with the desktop tools needed to make quick and inexpensive modifications. To stay competitive in today's rapidly changing call center environments demands easy and early adaptation.

The Single System Image View product is designed to make changes to applications painless. Modifications can be made even while an application is running. Users can quickly reshape applications to adapt to ever-changing business requirements.

EXPLOITING THE GRAPHICAL USER INTERFACE

A graphical user interface creates simple, intuitive representation of enterprise information. Graphical objects can be properly used to attract the eye to certain areas of a screen layout, visually directing the agent's attention. Each object has a specific behavior that is consistent between screens and even applications.

The combination of a good screen layout and consistent operating characteristics makes interaction with the system less demanding on agents. This provides a high degree of continuity that ultimately lessens application training time.

This new agent software exploits the graphical interface to create a friendly, interactive environment for agents. Distinct areas of the application windows are designed to contain certain information or functions that are consistently familiar, making new applications less intimidating.

Agent workstation software will offer users a complete set of predefined screen objects for handling a variety of data capture and retrieval methods. Using screen objects that are specifically designed for quickly selecting options or loading data fields will compress transaction time and diminish typing errors.

AGENT DIALOGUE MANAGEMENT

DIRECTING THE COURSE OF DIALOGUE

Normal attrition rates compel the typical call center manager to deal with constant agent turnover and training. At the same time, the quality of service provided to customers and prospects is expected to maintain a consistently high level that observes company policies and rules. Delivering a consistent and controlled business image requires directing the course of dialogue between agent and customer or prospect.

Through an application builder, the user should be able to build applications that establish or anticipate conversation flow. The application should handle both normal progression and exceptions, indirectly by application builder-embedded logic or directly by agent intervention. In telemarketing or telesales, this feature, in its most primitive form today, is referred to as scripting.

The Single System Image View product must be designed to broaden the concept of scripting into an industry-independent, total dialogue management of sound practice for any business application. It must help your agents accurately apply your business policies by presenting information in a focused, logical progression.

RESPONSE-DIRECTED NAVIGATION

Dialogue flow should be the underlying design principle of the application which encompasses business policies and procedures. These applications will use seemingly endless, response-based dialogue flows to provide the agent with the best path for interacting with the customer -- no matter how the customer directs the conversation.

INDIRECT NAVIGATION BASED ON INFORMATION RETRIEVAL AND COLLECTION

There may be more than one answer to a question and almost always there are exceptions to a rule. These are probably two of the most trying elements of application development. The application creator must be able to direct dialogue and navigation based on values retrieved from enterprise data sources, and from information collected during the conversation.

Logic and mathematics need to be easily embedded within the application, to test data field values and accordingly change dialogue flow. The history of previous conversations and transactions can also set the course. It's easy to imagine how this capability transcends industry and application borders as a tool for consistently controlling service quality and adherence to business policies.

MULTIPLE APPLICATION SUPPORT

Call center agents are often assigned to work more than one type of call transaction. Rather than designing monolithic host/server resident applications to handle this and other unforeseen situations, workstation-based client applications need to be deployed, each serving a different situation. Multiple applications must be ready to run, always in standby mode, at the agent's workstation. From the initiating call record, the workstation's client management environment should automatically select and launch the appropriate application on a call-by-call basis.

SEAMLESS INTEGRATION

As the next generation of desktop products proliferate into call centers, the true benefit of client/server applications will be realized. The Single System Image View concept has integrated access to multiple information sources. It will transform the multihost, serial and text-based information systems of today into the totally seamless, parallel access, multimedia, resolution-on-the spot solutions of tomorrow.

Figure 23
Next Generation Desktop Products

Through high speed networking/communication technologies, integration with legacy platforms becomes possible and economical. The Single System Image View combined with intelligent network technologies that are integrated with legacy systems makes us all winners. This solution helps protect investments in older technologies.

THE NEED FOR BETTER SWITCHING (PBX/ACD)

As discussed throughout this book, the outbound predictive dialing application gets us to the right customer or prospect as quickly as possible. By itself, predictive dialing has a limited life span. It serves you only from the moment the telephone number is being planned for dialing, until the desired person answers the phone, or a message is left.

While we have focused on how predicative dialing can connect you with your customers, the time spent once you're connected is even more important in achieving Customer Care. It's absolutely critical that the system be able to instantly (when the party says "Hello") bring up the customer's information, name, history, preferences etc., on the agent's screen. And at the same time and speed, bridge the agent's PBX/ACD-connected phone to that customer.

Most people in English speaking countries, when answering a telephone say "Hi," "Hello," or "Smith residence." If they hear no response, they may repeat the phrase again. If within 1-2 seconds, no sound is heard from the calling side, the called party will most likely hang up. This is why the predictive dialer's speed of connecting the agent to the person answering the call and displaying the pertinent information for the agent to begin speaking is so crucial. Any delays or hesitations can also set a poor tone for the conversation that follows.

Today's PBX/ACD switches are still not fast enough to handle a large number of predictive outbound calls.

PBX/ACD's were historically designed to process calls in the following order:

1. Incoming ACD calls
2. Incoming calls destined for switchboards or specific extensions
3. Outbound calls originated by various employees dialing from their desk telephones
4. The remaining types of calls, like those generated by predictive dialing applications using CTI, voice mail using data links, IVR, etc.

Therefore, calls generated by predictive dialing applications may be delayed by the switch while it processes inbound or outbound calls generated by other employees. For predictive dialing applications to achieve maximum efficiency and provide the best customer service, improved PBX/ACD products will emerge. They will assign predictive dialing applications the highest priority and at the same time provide better switching capabilities. But what's needed in the PBX/ACD to process the combined mission of inbound and outbound calls? Let's look at this next.

INBOUND/OUTBOUND CALL MANAGEMENT

Predicting inbound call traffic and allocating an appropriate staff to support it is a major challenge for an enterprise call center. Fortunately, technology is available that allows inbound agents to work as needed on outbound or inbound campaigns. This invention, Dynamic Inbound/Outbound, that we introduced to the market in 1989, has changed the entire call center staffing model.

This new innovation enabled designated agents to work simultaneously on both inbound and outbound campaigns, automatically, call by call, getting switched from one campaign to another as the needs of the call center dictate. This insures that the customer, whether calling in or being called, is immediately served by the first available agent best equipped to assist them.

This Dynamic Inbound/Outbound solution eliminates several costly needs:

- Overstaffing inbound campaigns for possible surges of incoming calls

- Placing customers and prospects on hold for long periods of time

- Dividing the business into inbound and outbound, with additional overheads of management, agents, communication equipment and other resources

- Assigning specific inbound or outbound related work to agents for a prolonged period of time

- Call center personnel needing to work from the same physical location

All agents can help with virtually all issues on-the-spot no matter where they are located. Because the Dynamic Inbound/Outbound solution analyzes every inbound/outbound call, it assures the fastest and best match between customer or prospect and agent.

The dynamic aspect of this solution when used with the Single System Image View agent workstation application, enables every designated agent to instantly accept a call, whether inbound or outbound. The dynamic algorithm ensures, on a call-by-call basis, that an agent may service an inbound call and, on conclusion, service an outbound call. And after that, another inbound or outbound call depending on real-time customer needs.

The two significant technologies that allow true nuisance-free Dynamic Inbound/Outbound call management are:

- Intelligent call cancellation of outbound calls before they are answered

- Dynamic Call Distribution™ (DCD) where the inbound calls and their priority level are weighed against outbound predictive/preview calls and Work Queue™ assignments

Just consider what happens if neither of these two technologies is deployed in the blending of inbound/outbound calls. Assume outbound predictive calls have been placed and are in the process of ringing in anticipation of available agents. Now assume a surge of incoming calls arrives and no inbound agent is yet available. If some of the already placed outbound calls are ringing and cannot be canceled, agents for which these calls are ringing cannot instantly be assigned to handle the incoming calls. This significantly increases the chance of having to put incoming calls on hold.

Lack of blended inbound/outbound call management capabilities on a call-by-call basis directly translates into customer dissatisfaction, because customers/prospects are more often put on hold and stay there longer. Combining outbound call cancellation with dynamic call distribution lets any outbound-assigned agent help with incoming calls. The regular outbound agent can be instantly connected to service the incoming caller. Similarly, regularly available inbound agents can be instantly assigned to service outbound calls when they are not needed for inbound. These combined technologies and services provide customers the highest level of service and businesses the lowest level of nuisance calls and operating costs — a win-win combination.

TECHNOLOGY TRENDS

FUTURE SWITCHING PLATFORMS

Though CTI technologies and applications are not a subject of this book, we must mention PBX/ACD based predictive dialing applications controlled via CTI. Until 1994, most predictive dialing solutions deployed across the globe have been based on proprietary CTI switching technologies provided by each predictive dialing supplier. These switches have been architecturally designed and specially tuned to connect, at extremely high speeds, the outgoing telephone line on which the call is being answered to the next available outbound agent.

Typically such connection time from the moment the call processing application software sends a command to the switch, to the time a connection is created, is approximately 20 to 50 milliseconds (0.02/0.05 seconds). This worst case individual switching time is very critical since, during high traffic conditions, many dozens of outbound as well as inbound calls must be processed by the switch. It is absolutely essential to be able to switch and connect, within a reasonable time frame, several outbound calls simultaneously to multiple agents (aggregating several worst case timings).

Consumers answering their telephone must be able to hear a live person as soon as they finish their "Hello." The switch infrastructure, in addition to inbound calls, must also be capable of handling hundreds of outbound calls, where at any time, a dozen or more consumers may be answering their telephones at the same time. The switch must further guarantee that the last such call will be connected in well under half a second.

No standards for even primitive predictive dialing Application Programming Interfaces (API's) were available from PBX/ACD suppliers until early 1992. This prevented many of us from deploying predictive dialing solutions using the industry's ACD platforms.

During '93 and early '94 a few of the key ACD suppliers announced their support for CTI based outbound call management. Again, no standards for CTI communications had been set. Only AT&T actually delivered its version of a call processing engine, the DEFINITY G3 digital switch equipped with Call Progress Detection and CTI interface. This introduction opened the door for developers to formally introduce predictive dialing applications integrated with an ACD inbound application using only one single switch as opposed to two -- one for inbound calls and one for predictive outbound calls.

Only PBX/ACD's that have complete functional Call Progress Detection (see Chapter 1) built-in and support high-speed CTI functionality, can be used as a single switching platform for managing both inbound and outbound predictive dialing call management.

It is absolutely critical that a PBX/ACD be consistently capable (via CTI link) of processing, routing and switching, at high speed, each voice call to an agent of choice.

This real-time, dynamic, work allocation process, where agents receive work based on externally controlled business applications, requires the PBX/ACD to be responsive to CTI commands for services on a call-by-call basis. Since large groups of agents dedicated to either inbound or outbound calls will no longer be valid in the Customer Care Center model of the future, the fundamental architecture and call routing algorithms in PBX/ACD's will need to change radically.

The new business model will inevitably often incorporate predictive dialing solutions in a variety of different ways. Predictive dialing is becoming a significant application in all business sectors on a global basis. When properly packaged, predictive dialing applications should fit into all inbound/outbound business practices, run on all significant operating systems and integrate all desktop platforms and applications.

As multitasking client/server architecture infiltrates diverse business segments, the predictive dialing application, through the use of API's, will become just another set of loadable modules, providing call processing services on demand.

TYING SWITCHING PLATFORMS TO DESKTOPS

Through various telephony-embedded applications, such as TSAPI (Telephony Services — API offered by Novell and AT&T and

Microsoft's TAPI), we have evidenced the birth of desktop telephony services, also called CTI desktop services.

These new services will increase productivity of both individual employees and corporate work groups promoting Customer Care in the everyday business tasks that drive the enterprise.

A multitude of telephony-based desktop applications are possible. External applications will tell the PBX/ACD to: keep dialing selected telephone numbers until it gets the desired person on the phone; route and transfer calls based on ANI, DNIS, time of day/week, or even spoken account number (using voice recognition technology); filter out user-specified unwanted calls, etc. Through the use of desktop applications and CTI telephony software, every employee and Care Center representative can define and organize their work-place environment on an individual basis.

Using CTI based protocols like AT&T's — ASAI, Northern Telecom's — Meridian Link, Intecom's — OAI, etc., and through packaged API's from IBM — CallPath, Novell — TSAPI, Dialogic — CT Connect, AT&T — Call Visor, etc., server- and desktop-based applications are being developed that take advantage of combined inbound/outbound predictive dialing and PBX/ACD functionality. Several switch manufacturers are planning to provide Call Progress Detection capabilities, facilitating the deployment of desktop automated call processing and predictive dialing applications.

Just imagine how every single employee could help service customers and improve efficiencies through the deployment of desktop CTI.

For example, employees can:

- Select the names of customers they need to call from their personal information manager and have the phone system automatically dial the numbers, utilizing the employee's time only when a live person answers the phone.
- Indicate to the phone system which incoming calls are important, and have the system automatically route these calls to wherever the employee is working when the call comes in (i.e., home, confer-

ence room, colleague's office, etc.).

- Filter out unwanted calls, either sending them to a voice mail system for retrieval later, or automatically forwarding them to any employee who would be better able to handle the call.
- Perform all current telephone functions (conferences, transfers, retrieving voice mail, etc.) using intuitive graphical based icons on their desktop PC, instead of telephone push buttons.

BACK INTO THE FUTURE

Looking back at the history of the PC and the PBX/ACD, we can actually predict what will happen in the future.

Remember what happened to the PC/desktop industry in the early '80s. It seemed like a revolution in home computing. Millions of PCs were sold, but only for a short burst of time. Sales leveled off when consumers realized there were few real, practical applications.

Looking at the PC market over the last 5 years, however, it seems like every store and retailer has been selling PCs on a massive scale. Statistics show that 50 million PCs are sold each year. Yes, practical, useful software applications are here, and consumers and businesses are benefiting immensely.

Let's see if we can learn from the PC evolution and predict what will happen to the PBX/ACD?

Look at the history of PBX/ACD switching platforms. At first, virtually every telecom supplier manufactured and/or sold PBX or key system (small, simple version of PBX). As businesses changed, consumers started taking more initiatives and created a new business era. The ACD inbound call processing era emerged. The call center was born. Call centers created the demand. Call center consumers wanted even more control of their destiny. So lots of functions and features were added to ACD platforms, including user-accessible databases. This enabled call center management to tune their operation by monitoring and reporting virtually any desired statistic retained in the database.

Some PBX/ACD manufacturers realized that call center management would not stop there — that eventually they would need full control over all calls, routing logic and more. So several manufacturers developed and marketed dumb switches with CTI embedded controls. Now call center management and their developers could design their own applications, logic and ACD environment.

But just like the early PCs, this dumb switch platform idea lasted only a short time.

Call center personnel realized that with the lack of standards for CTI, internal ACD functionality, response time, etc., their in-house development application would only work with a single, specific dumb switch. They would be stuck for years with the vendor that provided them the switch.

What may happen next?

If the market and subsequent technology follows trends we've identified and discussed, the following observations should hold true:

1. The dumb switch idea was not so dumb after all -- just too early.

2. Just like PCs today, tomorrow's switches will have to help process a variety of applications, and as such become dumb switch platforms controlled by external business applications.

3. Tomorrow's switches will probably become information routers/bridges helping funnel all types of information to/from any place in the world. This digital information will take many forms based on applications served.

4. Tomorrow's switches will have a standard set of functionality and next generation CTI-like controls. This will allow call center personnel and integrators to control their own Customer Care applications running with virtually any switch.

5. Tomorrow's switches will provide call processing and signal processing intelligence.

6. Tomorrow's switches will support (via business applications) call-by-call work queue assignment, routing voice and multimedia information to the best machine or human match.

7. Tomorrow's central offices will become intelligent information nodes to help form part of the super highway infrastructure web. This will allow people from any part of the world to communicate in any media type desired.

Not only is the new PBX/ACD architecture required to deliver total Customer Care in every aspect of a business, but the entire business process will mandate careful redesign so the customer always receives the highest level of attention and support. Whether the customer contact is by phone, fax, E-mail, video or face-to-face.

Figure 24
Total Customer Care

The dynamic nature of global business competitiveness brings with it this new paradigm — Customer Care. Real and global standards for CTI, Dynamic Inbound/Outbound, multimedia processing workflow, business reengineering methodologies, etc., will begin to emerge. Standards will allow businesses, call center management and integrators to rapidly build, and adapt to virtually all internal and external influences, their own Care Centers.

LOOKING FORWARD

Building platforms that are easily extendible and ready for the next stage is critical. The world is quickly moving into the next millennium where quantum leaps in technology will take place, impacting the way we live, work and play. The availability of information and the speed and ease with which it is accessed is already opening numerous doors of opportunity. We deeply believe in the applied information technologies that will enhance and enrich people's lives by improving the way they communicate and interact with each other.

In Five to Eight Years, Consider the Following (Within USA):

- 50% of consumers will be using Internet-like services such as E-mail, virtual shopping, banking/financial, notifications, voice communication, video and much more.
- 25% of consumers will be using CATV for information services (some similar to the above list, yet more in the area of on-demand entertainment, including voice and video communications).
- Telco's will be supplying TV/radio/broadcasting and entertainment-type programs in addition to global ISDN/ATM-based services.
- Local telco's will be supplying long distance and long distance suppliers providing local loop services. Cost of long distance will equal local calls. Billing will be measured based on the type of service and benefits it brings.
- In most cities, broadband communications wire line and wireless will be available on demand.
- Consumers will be loaded with choices. Customer services/support and ongoing relationships with consumers will be even more critical.
- Security will also be a key issue, so private/secured networks like Microsoft's NET, IBM Global Networks, ITS's, etc., will play a key role in enabling professional services across the world.
- We will be able to reach a person on their universal information device anywhere in the world. A single access number will be used to reach a person, even at work, but billing will be based on application types used (i.e. for home or business).

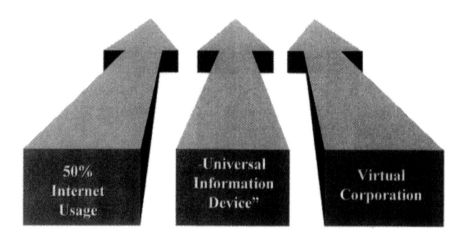

Figure 25
Looking Forward

- Business environment work queues will intelligently determine and match the best available resource (appliance or person) to address customer needs.
- Adaptive business methodology will evolve, allowing continuous business process evolution.
- Because of regulatory issues — pollution, traffic, hiring limitations — corporations will have multioffices and multicall centers (not only in major cities). Work-at-home and Work-on-the-go will become widespread.
- Centralized Information Services with decentralized personnel and other resources will be the corporate norm.
- The Virtual Corporation concept will become pervasive.
- Real-time resolutions of customer needs and real-time routing and assignment of all resources across a global enterprise will be the norm.

Figure 26
The Future

FINAL REMARKS

In the absence of a crystal ball, we rely on our imaginations and informed intuition to predict the future. One thing is sure — through applied technology, there will be tremendous opportunities to enhance our productivity and interactions as human beings.

Our success in creating and employing real business solutions is limited only by our willingness to embrace change. It's the key to fulfilling a vision of Customer Care and a better way of life for all mankind.

Volume and accessibility of information is approaching that of the air we breathe...available anywhere, at anytime, in whatever quantity we care to inhale. We must train ourselves to breathe deeply and develop strong lungs to successfully finish first in the race!

APPENDIX FIGURE A

Call Statistics Worksheet

Representative: _____   Supervisor: _____

Date: _____   Time Period _____ to _____

Activity for Each Call "TYPE"

Call #	Preview

Call "TYPE" in Seconds

RP	WP	NA	BZ	T/I	AM	VM	O

Activity for Each Call "TYPE"

WRAP

CALL TYPE EXPLANATION

RP = Right Party Contact
WP = Wrong Party Contact
BZ = Busy
NA = No Answer

T/I = Telephone Co. Recording
AM = Answering Machine
VM = Voice Mail Box
O = Any Other Condition

Step 1

- Record Preview Time for each call in the Preview Column (for the first call in a sequence, use the period of time elapsed from the point where the agent enters the command to bring up the record to the point where a connection of some type was reached ; NA, BZ,RP (etc.).

- It is important to note that Preview Time is not for each record, but for each dial attempt.

Step 2

- Record the length of time for the call "Type" occurred from the end of the dialing sequence until the end of the call (agent hangs up, client hangs up, etc.). There can only be one call "Type" per call.

- It is important to note that Preview Time is not for each record, but for each dial attempt.

Step 3

- Record the wrap time for each call beginning from when the agents hang up the phone to the point where they bring up the next record.

  You must enter a wrap time for each call 'Type' whether 0 seconds or 60 seconds.

Type of Calls	Number of Attempts	% Distribution	Time with Customer & Wrap Up (per call)	Manual Dialing Overhead (per call)	Weighted Average Transaction Time (per Call)
Right Party Contact					
Wrong Party Contact					
Answering Machine					
Totals				Overall Weighted Average	

APPENDIX FIGURE C

Type of Calls	Number of Attempts	% Distribution	Time with Customer & Wrap Up (per call)	Manual Dialing Overhead (per call)	Total Agent Time (per Call)
Right Party Contact					
Wrong Party Contact					
Answering Machine					
No Answer					
Busy					
Telephone Co.SIT Tones (3 Tones)					
Other Conditions (High & Dry, fax, modem)					
Totals					

INDEX

A

T

U

ALEKSANDER SZLAM A BIOGRAPHY

Aleksander Szlam, the founder, chairman and CEO of Melita International Corporation has lead and guided the company from its startup to its becoming a world leader and provider of Call Center solutions. From the beginning his personal vision has been to provide solutions that enhance communications and benefit people around the world.

Throughout Mr. Szlam's 20 year career, he has been a driving force behind establishing Call Center industry trends and standards and building solutions based on innovative technologies and human factors. The Customer Care™ Center philosophy that he also introduced in 1993 has been gaining worldwide momentum.

Mr. Szlam has collected over 30 world patents for his technological inventions and concepts. Along with inventing the predictive dialer and the synchronized voice and data "screen pop," he is also credited as being the first to introduce: personal computers as workstations for outbound Call Centers, local area networks for predictive dialing solutions, CTI for PBX/ACD integrated predictive dialing, and Dynamic Inbound/Outbound call management. Many of these inventions have been deployed worldwide as an integral part of Call Center solutions offered by providers such as: AT&T, Northern Telecom, Alcatel, IBM, Digital Equipment Corporation, Sprint, MCI, British Telecom and Dialogic.

For his efforts as an inventor and entrepreneur, Mr. Szlam was named Inc. Magazine's Entrepreneur of the Year, Southern Region in 1991, while his company has repeatedly made Atlanta's prestigious Fast Tech 50 list of growing technology companies. Today, Mr. Szlam is focusing on formulating and executing long term strategic product plans that will take Melita to the year 2000 and beyond.

Mr. Szlam earned his Bachelor's and Master's degrees in electrical engineering (with emphasis on digital signal processing), from the Georgia Institute of Technology graduating with honors. Prior to starting Melita, he worked as a design engineer and scientist at companies such as Lockheed Corporation, NCR and Solid State.

The company that he founded and incorporated in 1979, Melita International, formally opened its doors in 1983. The company's first product, built in 1979, was an autodialing and call receiving system commissioned by Wisconsin Public Service Corporation of Green Bay. Installation of Melita's autodialer allowed the power company to dispatch repairmen during emergencies and, at the same time, dramatically cut operation costs.

Melita products are installed on over 25,000 DOS, Windows, Macintosh, and OS2 desktops at 1500 installations in 17 countries on six continents. Melita call processing solutions are found in Call Centers in many Fortune 500 companies, including some of the world's largest banks and collection agencies, large telemarketing centers, telco's, cellular providers, service bureaus, utility companies, universities, healthcare, cable companies, retailers and other industries.

Melita, with numerous offices in North America and international offices in Singapore, United Kingdom and Colombia, has established sales and support partners in many world regions.

KEN THATCHER A BIOGRAPHY

ATLANTA, GA – Currently, Mr. Thatcher is Senior Vice President, Professional Services. During his twelve years with Melita, he has served as Senior Vice President and Managing Director, Melita Europe, and Senior Vice President, Marketing and Product Development.

Throughout Mr. Thatcher's career in the computer industry, he has held senior management positions at UNIVAC and Honeywell. His experience with Call Center technology, as it relates to systems, solutions, design, implementation, and CTI technology provides Mr. Thatcher with a broad knowledge of the computer and telecommunications industry.

As a result of his 40-year career in the computer industry, Mr. Thatcher is frequently invited to speak throughout Europe and the United States where he has given presentations at conferences, user group meetings and seminars.

He is a native of Indiana, holds a bachelor's degree in mathematics from the University of Illinois, and attended Emory University's Graduate School of Business.